Lecture Notes
in Business Information Processing   205

Series Editors

Wil van der Aalst
Eindhoven Technical University, Eindhoven, The Netherlands
John Mylopoulos
University of Trento, Povo, Italy
Michael Rosemann
Queensland University of Technology, Brisbane, QLD, Australia
Michael J. Shaw
University of Illinois, Urbana-Champaign, IL, USA
Clemens Szyperski
Microsoft Research, Redmond, WA, USA

More information about this series at http://www.springer.com/series/7911

Esteban Zimányi · Ralf-Detlef Kutsche (Eds.)

Business Intelligence

4th European Summer School, eBISS 2014
Berlin, Germany, July 6–11, 2014
Tutorial Lectures

 Springer

Editors
Esteban Zimányi
Université Libre de Bruxelles
Brussels
Belgium

Ralf-Detlef Kutsche
Technische Universtät Berlin
Berlin
Germany

ISSN 1865-1348 ISSN 1865-1356 (electronic)
Lecture Notes in Business Information Processing
ISBN 978-3-319-17550-8 ISBN 978-3-319-17551-5 (eBook)
DOI 10.1007/978-3-319-17551-5

Library of Congress Control Number: 2014934656

Springer Cham Heidelberg New York Dordrecht London

Printed on acid-free paper

Springer International Publishing AG Switzerland is part of Springer Science+Business Media
(www.springer.com)

Preface

The Fourth European Business Intelligence Summer School (eBISS 2014) took place in Berlin, Germany, in July 2014. Tutorials were given by renowned experts and covered several recent topics in business intelligence. This volume contains the lecture notes of the summer school.

The first chapter surveys the domain of requirements engineering for decision support systems. This is done in the context of a real-world application for analyzing the impact of the Chagas disease. This disease is classified as a life-threatening disease by the World Health Organization (WHO) and causes numerous deaths every year. The development of the Chagas Information Database (CID) is part of WHO's strategy for advancing in the disease control. CID is a decision support system to support national and international authorities in both their day-by-day and long-term decision making. The paper describes the results of applying Pohl's Framework for the requirements engineering phase of this project.

The second chapter presents the application of visual analytics for enabling a multiperspective analysis of mobile phone call data records. The analysis of human mobility is a hot research topic in data mining, geographic information science, and visual analytics. While a wide variety of methods and tools are available, it is still hard to systematically consider a dataset from multiple perspectives. The paper presents a workflow that enables a comprehensive analysis of a publicly available dataset about mobile phone calls of a large population over a long time period. The paper concludes by outlining potential applications of the proposed method.

The third chapter gives an overview of how the Web of documents has evolved into what is referred to as Linked Data. The paper starts with a description of the evolution that led from the first version of the Web to the Web of data, sometimes referred to as the Semantic Web. Based on the "data > information > knowledge" hierarchy, the article makes explicit the structures of knowledge representation and the building blocks of the Web of data. Then, the paper shows how RDF (Resource Description Framework) data can be managed and queried. After that, the paper delves on ontologies, from their creation to their alignment and reasoning. The article concludes by pointing out research and development perspectives in the linked data environment.

The fourth chapter gives a survey of supervised classification on data streams. Research in the statistical learning and data mining fields in the last decade resulted in many learning algorithms that are fast and automatic. However, a strong hypothesis made by these learning algorithms is that all examples can be loaded into memory. Recently, new use cases generating huge amounts of data have appeared, such as monitoring of telecommunication networks, user modeling in dynamic social networks, web mining, etc. As the volume of data increases rapidly, it is now necessary to use incremental learning algorithms on data streams. The article presents the main approaches of incremental supervised classification available in the literature.

Finally, the fifth chapter presents a survey of existing techniques of knowledge reuse and provides a classification approach for them. The importance of managing organizational knowledge for enterprises has been recognized since decades. Indeed, a systematic development and reuse of knowledge will help to improve the competitiveness of an enterprise. The paper investigates different approaches for knowledge reuse from computer science and business information systems. It proposes a classification approach for these techniques based on the following criteria: reuse technique, reuse situation, capacity of knowledge representation, addressee of knowledge, validation status, scope, and phase of solution development.

In addition to the lectures corresponding to the chapters described above, the eBISS 2014 had three other lectures, as follows:

– Frithjof Dau, from SAP Research, Germany: CUBIST - Combining and Uniting Business Intelligence with Semantic Technologies
– Asterios Katsifodimos, from Technical Universität Berlin, Germany: Big Data looks tiny from Stratosphere
– Roel J. Wieringa, from University of Twente, The Netherlands: Design Science Methodology for Business Intelligence

These lectures have no associated chapter in this volume, because their content is reported in recent publications, respectively, [1], [2], and [3].

We would like to thank the attendants of the summer school for their active participation, as well as the speakers and their co-authors for the high quality of their contribution in a constant evolving and highly competitive domain. Finally, the lectures in this volume greatly benefit from the comments of the external reviewers.

January 2015   Esteban Zimányi
                                                                            Ralf-Detlef Kutsche

References

1. Dau, F., Andrews, S.: Combining business intelligence with semantic technologies: the CUBIST project. In: Hernandez, N., Jäschke, R., Croitoru, M. (eds.) ICCS 2014. LNCS (LNAI), vol. 8577, pp. 281–286. Springer, Heidelberg (2014)
2. Alexandrov, A., et al.: The stratosphere platform for big data analytics. VLDB J. 23(6), 939–964 (2014)
3. Wieringa, R.J.: Design Science Methodology for Information Systems and Software Engineering. Springer, Heidelberg (2014)

Organization

The Fourth European Business Intelligence Summer School (eBISS 2014) was organized by the Department of Computer and Decision Engineering (CoDE) of the Université Libre de Bruxelles, Belgium, and the Database Systems and Information Management (DIMA) Group of the Technische Universität Berlin, Germany.

Program Commitee

Alberto Abelló	Universitat Politécnica de Catalunya, BarcelonaTech, Spain
Marie-Aude Aufaure	École Centrale Paris, France
Ralf-Detlef Kutsche	Technische Universität Berlin, Germany
Patrick Marcel	Université François Rabelais de Tours, France
Esteban Zimányi	Université Libre de Bruxelles, Belgium

External Referees

Waqas Ahmed	Université Libre de Bruxelles, Belgium
Catherine Faron Zucker	Université Nice Sophia Antipolis, France
Dilshod Ibragimov	Université Libre de Bruxelles, Belgium
Marite Kirikova	Riga Technical University, Latvia
Georg Krempl	Otto-von-Guericke University Magdeburg, Germany
Pascale Kuntz	Laboratoire d'Informatique de Nantes Atlantique, France
Gerasimos Marketos	University of Piraeus, Greece
Faisal Orakzai	Université Libre de Bruxelles, Belgium

Contents

On the Complexity of Requirements Engineering for Decision-Support Systems: The CID Case Study

Ruth Raventós, Stephany García, Oscar Romero[(✉)],
Alberto Abelló, and Jaume Viñas

Department of Service and Information Systems Engineering,
Universitat Politècnica de Catalunya, Barcelona, Spain
{raventos,sgarcia,oromero,aabello,jvinas}@essi.upc.edu

Abstract. The Chagas disease is classified as a life-threatening disease by the World Health Organization (WHO) and is currently causing death to 534,000 people every year. In order to advance with the disease control, the WHO presented a strategy that included the development of the Chagas Information Database (CID) for surveillance to raise awareness about Chagas. CID is defined as a decision-support system to support national and international authorities in both their day-by-day and long-term decision making. The requirements engineering to develop this project was particularly complex and Pohl's framework was followed. This paper describes the results of applying the framework in this project. Thus, it focuses on the requirements engineering stage. The difficulties found motivated the further study and analysis of the complexity of requirements engineering in decision-support systems and the feasibility of using said framework.

Keywords: Decision-support systems · Business intelligence · Requirements engineering

1 Introduction

The World Health Organization (WHO)[1], a United Nations (UN) agency founded in 1948, is the directing authority for health matters, responsible to provide leadership on global health matters by setting standards and providing technical support to monitor the health trends. In October 2010, the WHO launched the First Report on Neglected Tropical Diseases (NTD) [1], which included the "Chagas disease" among others. Chagas Disease (also known as Human American Trypanosomiasis) is classified as a life-threatening disease caused by a parasite named "protozoan parasite". According to the report, Chagas is nowadays usually found in 21 Latin American countries, where the disease is transmitted to humans by the faeces of infected triatome bugs. The WHO report has estimated

---

[1] http://www.who.int/about/en/.

© Springer International Publishing Switzerland 2015
E. Zimányi and R.-D. Kutsche (Eds.): eBISS 2014, LNBIP 205, pp. 1–38, 2015.
DOI: 10.1007/978-3-319-17551-5_1

seven to eight million people infected worldwide, although the concern is that the number of infected people keeps increasing year by year.

By being categorized as NTD, not much attention is drawn by the Chagas disease. In fact, worldwide population is not aware that statistics indicate that NTDs have already caused substantial illness for more than one billion people of our world's poorest countries. Due to the worrying situation, in 2010 the World Health Assembly (WHA) decided to approve the "Chagas disease: control and elimination" resolution (SDC10) [1]. But it was not until 2013 that the Tricycle Strategy of the Programme on Control of Chagas disease, was presented (VN) [28]. The presented strategy had the objective to advance in the disease control by creating a Chagas Information Database (CID) system for surveillance to raise awareness on the Chagas disease. In particular to facilitate the access to information from different sources related to Chagas disease, to then exploit the data by means of dashboards, and analytical tools (e.g., visualize disease statistics, maps and diagrams for transmission routes; infested patients).

The CID project involved specialized stakeholders from very different domains and organizations such as health specialists and national authorities from (or linked to) WHO, entomologists from different research centers of Argentina and software analysts from Universitat Politècnica de Catalunya (UPC-BarcelonaTech), who had complementary yet non-trivially integrable perspectives of the Chagas disease. For instance, entomologists set their focus on studying the triatome bugs (the main disease vector) and how to interrupt this transmission by such vectors. Doctors, in turn, focused on patient diagnosis and treatment and the WHO monitors all relevant aspects of the disease, including provided infrastructures and drug delivery in each country, and provided the multi-perspective expertise needed to glue all of them together. In consequence to this, software analysts were expected to deal with the highly-complex domain of the Chagas disease to begin collecting the requirements before creating the CID system.

From the very beginning, the project was acknowledged as a highly complex one. First of all, the WHO Programme did not have a pre-determined or defined proposal of the information and surveillance system they needed. In addition, the interdisciplinary knowledge among the principal stakeholders doctors, entomologists, requirement analysts and database experts - to gather together to reach consensus on the crucial project issues was a true challenge. Several meetings were held in order to let business and technical analysts understand the basics of Chagas: i.e., diagnostic, frequently used drugs, vector control and health systems.

In addition to the understanding among the different stakeholders mentioned above, some other problems were found, mainly related to data, during the requirements engineering process.

In order to achieve the goal of eradicating Chagas (CID's ultimate goal), we needed to integrate data corresponding to the different facets (i.e., medical, vectorial, etc.) of the Chagas disease. Its main difficulty came from the heterogeneity (either in the storage system or semantics) in the multiple data sources and how to cross-analyze all such information adding the geographical and temporal dimensions and treat them at different granuralities (of analysis).

Even though the Chagas disease initially appeared in South America today, due to migratory flows, it is scattered around the whole world. Thus, information can come from any country and users can speak any language. This clearly raises a linguistic problem that had to be overcome by a multilingual interface.

Another issue was the quality of data. To decide what mechanisms would be put in place to check as many constraints as possible to ensure the quality of the data introduced.

As mentioned, the CID project is primarily a DSS. Thus, all relevant data related to Chagas must be gathered, properly stored and managed (this means integrated and cleaned) in order to allow the WHO to analyse and cross data from disparate, yet relevant, from different points of view.

Therefore, in order to systematically tackle the complexity of the project, during the requirements engineering phase we followed K. Pohl's requirements engineering framework [2] to understand, agree, and document all the requirements of the system to be built in compliance with the relevant documentation formats and rules.

This framework has become a well-established for requirements engineering mainly because of the following reasons:

- The first and main reason was that the framework defines the major structural blocks and elements of a requirements engineering process (e.g., Elicitation, Negotiation, Goals, etc.).
- It provides a comprehensible and well-structured base for the fundamentals, principles and techniques of requirements engineering.
- It is not adhered to any specific methodology, neither to any type of software project.
- In addition, this framework consolidates various research results and has been proven to be successful in a number of organizations for structuring their requirements engineering process.

The rest of the paper describes the requirements engineering process of the CID project by applying Pohl's framework emphasizing the complexity found.

The importance of this project, in addition to its own purpose, is that many of the lessons learnt can be mapped to most decisional systems. For this reason, this case study has been the base of further research about requirements engineering for decision-support systems and has been the reason for further research conducted in this area in our research group to propose a new framework for requirements engineering of decision-support systems, which is nowadays missing. As previously discussed, to cover this gap we followed Pohl's requirement engineering framework and here we show by example how each of its constructs can be adapted to decisional systems. All in all, this paper is an excellent case study showing all the complexities found at this stage when developing a DSS.

2   Requirements Engineering of the CID Project

Every single project must begin with the statement of a requirement. In computer science, a requirements refers to the description of how a software product should perform [3]. The IEEE 610.12-1990 [4] defines "requirement" as follows:

1. "A condition or capability needed by a user to solve a problem or achieve an objective"
2. "A condition or capability that must be met or possessed by the system or system component to satisfy a contract, standard, specification, or other formally imposed document"
3. "A documented representation of a condition or capability as in (1) or (2)"

The CID project was promoted by the World Health Organization (WHO) to change the current reality of a neglected disease, Chagas. The initial vision was to develop a software system that would help controlling and managing all the relevant information related to the disease, i.e., detection, treatment, drug availability / delivery, transmission interruption, normative, etc. so, it would be used as support to decide what actions to take to combat the disease.

More precisely, the main objective of this project is to design a data warehouse for the global WHO information and surveillance system to control/eliminate Chagas disease, which should be based on the following components (to be gathered world-wide):

– Current legal, normative, political and economic frameworks about Chagas.
– Information and surveillance systems in place (with information on circulating parasites, presence of infected and non-infected vector insects, hosts and humans).
– Healthcare data (including screening, diagnosis and care measures and coverage, through biomedical and psychosocial approaches).
– Chagas disease affected population (including population at risk of being infected, estimated number of infected and ill people, already diagnosed and treated Chagas disease patients, from congenitally infected newborns to adults)
– Transmission routes (active and interrupted transmission routes and implemented measures to prevent and interrupt them).

According to main objective of the project, it was necessary to perform a specific group of tasks to analyse the needs, goals, stakeholders and context of the required system. It meant to carry out the following tasks:

– The state of the art related to the system. Is there any other similar system out there?
– Data sources from where to extract the needed data (quality and risks).
– Scenarios, use cases and boundary of the system to be built.
– Software requirements (functional and quality requirements) and constraints.
– Conceptual model of the system.
– Estimation of the timing and budget of the project.

Consequently, the requirements engineering stage was divided in three main tasks (in parenthesis, the amount of hours devoted to each task): state of the art (90 h), definition of actors, use cases and system sequence diagrams (150 h) and definition of the conceptual model (100 h). The project spanned for 3 months.

The main distinction of defining the requirements and developing the design of a system is that the requirements focus on "what" the system should do,

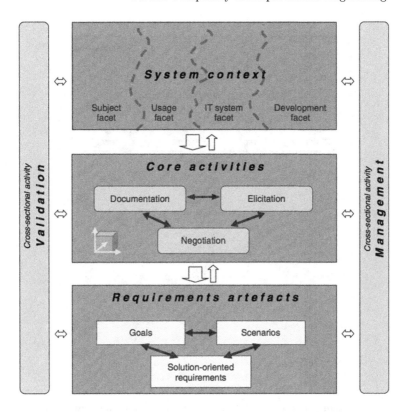

Fig. 1. Pohl's requirements engineering framework [2]

rather than "how" it should be done [3]. Meaning that the requirements are centered on defining the problem "what should be developed", while the design specifies "how the system should be developed" [2].

Therefore, requirements engineering is a cooperative, iterative, and incremental process which aims to ensure the following [2]:

1. All relevant requirements are explicitly known and understood at the required level of detail.
2. Sufficient agreement about the system requirements is achieved between the stakeholders and the IT team involved.
3. All requirements are documented and specified in compliance with the relevant documentation/specification formats and rules.

Pohl's requirements engineering framework (see Fig. 1) consists of three main building blocks and two cross-sectional activities which are explained in the following sections of this document.

3   System Context

The requirements in every software system are heavily influenced by the system context. Consequently the requirements for these systems may not be defined

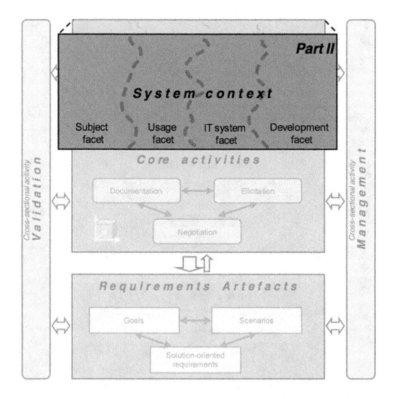

Fig. 2. System context in Pohl's requirements engineering framework [2]

without considering the environment where the system is embedded. These include material or immaterial objects such as the following: technical or non-technical systems, people, technologies, business processes and work-flows, laws, already existing hardware and software components, other systems that will interact with the new system, physical laws, safety standards, system users, etc. Therefore, the understanding of the context is a crucial preparation in order to develop a successful requirements specification.

The system context is structured in the following four different context facets as shown in Fig. 2 which must be considered as the prerequisite phase on the Requirements Engineering process:

1. **Subject Facet**
   This facet is composed mainly of the objects and the events that must be represented in the system; given that the system must process or store information about these. Additionally, this facet also includes aspects that constraint the representation of the system such as data privacy laws disallowing the storage of a certain type of data or accuracy constraints.

2. **Usage Facet**
   This facet which refers to all aspects concerning the usage of the system by people or other systems such as: user groups with specific characteristics,

usage goals, desired usage work-flows, or modes to interact with the system through the system's interface.

3. **IT Facet**
   This facet encompasses all aspects related to the operational or technical environment in which the system is deployed. IT facet includes hardware and software components (e.g., software platforms, networks, existing components), IT strategies and policies.

4. **Development Facet**
   Last but not least, the development facet encompasses all the aspects which influence the system's development process, for instance: process guidelines, development tools, quality assurance methods, and other techniques to assure quality in the system. This may always be restricted by law, for instance if there are only certain tools the client requested to be used during the systems' development.

3.1   CID Project Subject Facet

To elicit the information that should be represented in the CID system, first, we needed to understand what is the Chagas disease, its distribution around the world, the signs and symptoms, the transmission, treatment, control and prevention of the disease and its relation with AIDS. All the aspects presented in this section are of extreme relevance to support the WHO in its ultimate objective of eradicate Chagas.

The Disease. Chagas disease[2] (also known as Human American Trypanosomiasis) is a potentially lifethreatening illness caused by the protozoan parasite, *Trypanosoma cruzi*. It is found mainly in endemic areas of 21 Latin American countries, where it is mostly transmitted to humans by the faeces of infected triatome bugs. At present about 7 to 8 million people worldwide are estimated to be infected with *T. cruzi*, including disease non-endemic countries where Chagas disease has been spread through population movements, mainly migration. Chagas disease and its causal agent, *T. cruzi*, have existed for millions years in the Americas in sylvatic cycles (enzootic disease[3]). Humans arrived in the Americas at least between 26,000 and 12,000 year ago, either overland or across the oceans. Settled communities started to implement agricultural activities and domestication of animals. Over the past 200–300 years, deforestation because of agriculture, livestock rearing and the opening of transportation routes (railroads and highways), cause that Triatoma vectors gradually lost their primary food source: wild-animal blood. Vectors[4] adopted the areas surrounding human

---

[2] Chagas disease is named after Carlos Ribeiro Justino Chagas, a Brazilian doctor who first discovered the disease in 1909 [5].

[3] A disease of wild animals [5].

[4] Vector that have been transmitting the Chagas disease among animals for 10 million years [5].

dwellings and started to feed on the blood of domestic animals and humans. A new cycle was established and Chagas disease became a zoonosis[5]. Endemic Chagas disease in Latin America began as neglected disease of poor, rural and forgotten human populations. Wild triatomines progressively adapted to the domestic environment, living in cracks in the mud walls and roof of huts. In the second half of the XX century Chagas disease was recognized as an important urban medical and social problem because of progressive urbanization of the rural population in Latin America. In 1960, a WHO Expert Committee estimated that about 7 to 8 million people per year were infected due to blood transmission in Latin America [5].

Distribution. Chagas disease occurs mainly in the following 21 Latin America countries: Argentina, Belize, the Bolivarian Republic of Venezuela, Brazil, Chile, Colombia, Costa Rica, Ecuador, El Salvador, French Guyana, Guatemala, Guyana, Honduras, Mexico, Nicaragua, Panama, Paraguay, Peru, the Plurinational State of Bolivia, Suriname and Uruguay, but no cases have been detected so far in Caribbean countries. Additionally, in the past decades Chagas disease has increased the spread worldwide and it is now detected in the United States of America, Canada, 17 European countries (mainly in Belgium, France, Italy, Spain, Switzerland and the United Kingdom, but also in Austria, Croatia, Denmark, Finland, Germany, Luxembourg, the Netherlands, Norway, Portugal, Romania and Sweden) and some Western Pacific countries, mainly Australia and Japan, see Fig. 3. The disease presence outside disease endemic countries is mainly due to population mobility between Latin America and the rest of the world, but less frequently due to infection through blood transfusion, congenital transmission (from infected mother to child) and even organ transplantation, laboratory accident and child adoption [5].

The above-mentioned population movements have created new epidemiological, economic, social and politic challenges for rich and poor countries where *T. cruzi* has been spread [5]. In countries with absence of insect vectors, such the European ones, main threats come from infected blood transfusions, organ transplantations or congenital transmission. On the other hand, the number of cases of infection among people travelling to Latin America for tourism or work reasons has been increasingly reported. Therefore, these countries are progressively realizing that they need to improve their information and surveillance systems at national and supranational levels, increase detection and care of patients with Chagas disease, implement additional controls for blood banks and organ transplantation, and include the differential diagnosis of Chagas disease within travel medicine.

Signs and Symptoms. Chagas disease presents itself in two successive phases, an acute phase and a chronic phase.

1. Acute phase: The acute phase lasts for about two months after infection. Historically, with high vector infestation of dwellings and consequent intradomiciliary transmission, the majority of cases were detected before the age of

---
[5] A disease that is transmitted between animals and humans endemically [5].

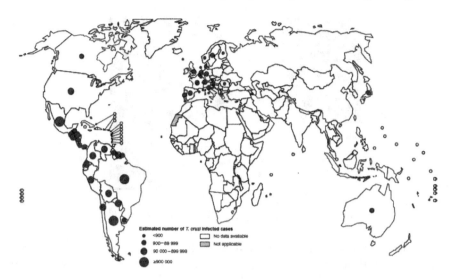

Fig. 3. Global distribution of cases of Chagas disease, based on official estimates, 2006–2010 [5]

15 years, with the highest frequency between the ages of 1 and 5 years [5]. Acute Chagas disease, however, can occur at any age and at present, for instance, an important number of cases are related to oral transmission in territories such as the Amazon basin or humid Andes. During the acute phase, a high number of parasites circulate in the blood. Most *T. cruzi* infected people present mild or absent symptoms, but can also have fever, headache, enlarged lymph glands, pallor, muscle pain, difficulty in breathing, swelling and abdominal or chest pain. And despite most acute infections are unrecognised [5], in less than 50 % of people bitten by a triatomine bug, characteristic first visible signs can be a skin lesion or a purplish swelling of the lids of one eye, Fig. 4.

In most cases, congenital transmission, produced by the transmission of *T. cruzi* parasites from the infected mother to the child, occurs at the end of the pregnancy or delivery and is a specific acute case. Most infected infants are asymptomatic and have normal weight and vital signs, but some, mainly those born prematurely, may show clinical manifestations, such as low body weight, jaundice, anaemia, hepatomegaly or ocular lesions, among others [5].

2. Chronic phase: The chronic phase of Chagas disease starts when the parasite falls to undetectable levels and general symptoms and clinical or manifestation are initially not present. During the chronic phase, the parasites are hidden mainly in target tissues: the heart and digestive muscle. Current knowledge on all changes produced during the chronic phase is still incomplete, and this initial asymptomatic period of the chronic phase is called indeterminate or latent form. About 50 %–60 % of infected people will remain in this form lifelong [5]. On the other hand, up to 40 % of chronically infected individuals will develop a cardiac, digestive, neurological or mixed form. Cardiac manifestations are

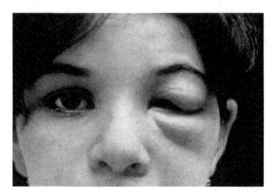

Fig. 4. Sign in acute phase of Chagas disease [6]

the most important clinical consequence of the *T. cruzi* infection. Up to 30 % of patients will suffer from cardiac disorders. These forms occur mainly 10–20 years after the acute phase of the disease and include a broad range of type of damage and clinical manifestations, from mild symptoms to heart failure, arrhythmias and sudden death [5]. The digestive form of Chagas disease has been described in populations south to the Equator line. The infection can lead to dysfunction of the digestive system with dilation of oesophagus and colon (megaoesophagus). Up to 10 % suffer from digestive, neurological or mixed alterations [5].

Transmission. One of the most important challenges to control/eliminate the Chagas disease is to interrupt its transmission. Chagas disease can be transmitted by six different transmission routes, mainly vectorial, oral, congenital (mother to child) and blood tranfusion routes, but also by organ transplantation and laboratory accidents. The vectorial transmission route has been the most important one. It is mainly found in the 21 endemic countries of Latin America, where most of the insects infected with *T. cruzi* are found. These bugs, which typically live in the cracks of poorly-constructed homes of rural or suburban areas, normally hide during the day and feed on human blood when the night falls [5]. In fact, *T. cruzi* parasites are mainly transmitted by the contact with faeces and urine of blood-sucking triatomine bugs. The itching caused by the bits leads humans scratching or putting in contact infected faeces with the bite wound, other skin lesions or mucous membranes (mainly in the mouth and eyes), allowing the parasites to enter the circulation system [5]. Another vector-related route of transmission is the oral or food-borne transmission, through the ingestion of food contaminated with triatomine bugs faeces, very prevalent in territories such as the Amazon basin. The other transmission routes *T. cruzi* can also be transmitted by [5]:

– Blood transfusions using blood from infected donors.
– Vertical transmission (mother to child).

- Organ transplants using organs from infected donors.
- Laboratory accidents.

Treatment. In order to kill the parasite, Chagas disease can be treated with either benznidazole [7] or nifurtimox [8]. Both medicines are almost 100 % effective in curing the disease if given soon after the infection at the beginning of the acute phase. However, the efficacy of both reduces the longer a person has been infected. Treatment is also indicated for those in whom the infection has been reactivated due to immunedepression, for infants with congenital infection and for patients during the early chronic phase [5]. Treatment should be offered to infected adults, especially those with no symptoms. Anyway, the benefits of medication in preventing or delaying the development of Chagas disease should be weighed against the long duration of treatment and possible adverse reactions [5]. Benznidazole produces adverse reactions in 20 %–40 % of adult patients, but they usually disappear when the treatment is finished. In a few cases treatment has to be discontinued because of side-effects. On the other hand, children have a higher tolerance of the treatment [1]. Additionally, pregnant women or people with kidney or liver failure should not medicate with neither both benznidazole nor nifurtimox. And nifurtimox is also contraindicated for people who suffer neurological or psychiatric disorders [1].

Control and Prevention. There is no vaccine to prevent from *T. cruzi* infection. Depending on the region, vector control is the most effective method of preventing Chagas disease. Blood screening is necessary to avoid infection through blood and blood products transfusion or organ transplantation. WHO recommends the following approaches to prevention and control [1]:

- Insecticide spraying of houses and surrounding areas.
- House improvements to prevent vector infestation (such as plastered walls, cement floors or corrugated-iron roofs).
- Personal preventive measures such as bed-nets.
- Good hygiene practices in food preparation, transportation, storage and consumption.
- Screening blood donors.
- Testing of organ, tissue or cell donors and receivers.
- Screening of newborns and other children of infected mothers to provide early diagnosis and treatment.
- Use of safety standard protocols to prevent laboratory accidents, especially when dealing with cultures and trypomastigote forms [1]

Chagas Disease and AIDS. Reactivation of Chagas disease in chronically infected persons has been associated with immunodeficiency states, such as those related to hematologic malignancies, organ transplantation and corticosteroid therapy [8]. Since the 1980 s, reactivation of Chagas disease has also been observed in patients with human immunodeficiency virus (HIV) infection.

The increased phenomena of population mobility in the last decades, including urbanization and migration to other continents, has permitted the geographical overlap of both HIV and *T. cruzi* infections, either in endemic and non-endemic countries [9].

3.2 CID Project Usage Facet

Chagas disease initially appeared in South America. However, due to migratory flows, today it is scattered around the whole world. Thus, information can come from any country and users can speak any language. This clearly raises a linguistic problem, and consequently a multilingual interface must be offered, which, from the ICT point of view, is a well know problem that can be easily solved. Nevertheless, two not so obvious problems are also relevant to our case.

Of course, the main source of information will be the governments of the different countries. However, the collaboration of researchers and NGOs (or simply concerned citizens) is also expected. Thus, we need to keep track of who introduced what in the system (also known as data provenance). This will guarantee that we can trace back to the original source any chart or figure. Nevertheless, this is not enough. It is necessary to provide some kind of quality guarantees for any unofficial source. Any data introduced must be backed by some kind of documentation (e.g., journal publication of scientific results). Then, a group of experts designed by WHO will validate the information.

Additionally, such kind of sensitive epidemiological data can easily raise concerns in the governments. Therefore, two different versions of the system will be produced: a centralized Web interface, and a desktop application. This will allow the off-line introduction of data, that after the necessary checking by the national authorities can later on be uploaded in the centralized WHO server. For the data in the server, under the responsibility of WHO, the necessary security mechanisms must be provided in order to guarantee confidentiality.

Therefore the critical stakeholders identified (along with their usage goals) were the following:

– World Health Organization: promoter of the project interested in visualize, manage and analyse all the information related to Chagas. The WHO specialists will be the ultimate decision makers and need all relevant data, coming from any country, non-profit organization or research group in order to gain a clear picture of the current status of the disease. According to such picture, the WHO will establish a worldwide roadmap that, again, thanks to the system, will be monitored, analysed (see what impact had) and define the next milestones. All in all, following the natural BI cycle.
– Health Ministry Officers: responsible to provide, modify and delete any Chagas information (about healthcare, transmission interruption, systemic or normative information) related to their respective countries. These stakeholders will provide relevant data about their countries related to Chagas but they will also have access to reports and analysis created for their own uploaded data to let them gain insight of the situation in their countries (e.g., are they meeting the milestones established in the WHO roadmap?).

- Researchers / NGOs: interested in provide information related to healthcare and transmission interruption and also in the visualization and analysis of such information. The ultimate goal of these stakeholders is to perform in-depth analysis by means of developing complex models to simulate and predict certain behaviours. For example, according to climate data, where could the kiss bug be endemically found in a time frame of 5 years?
- Technical groups: who take care of the consistency of the information provided by other stakeholders. They are as well considered as stakeholders because technical groups will continuously monitor the current status of the disease by performing random analysis over the data. For example, one important contribution of such groups are estimations published (see Sect. 3.3). The difference with the WHO stakeholder is that these groups are supporting / consultant groups of experts for the WHO, whose experts have the last word about the roadmap to be develop for the next years.

Above, the stakeholders are classified according to their position. Below, we alternatively classify them from the point of view of how they would introduce, manage, delete, query, analyse or visualize information with the system as follows (note that a position -e.g., researchers- may play several of these roles below):

- Data producers that will introduce relevant information in the system. Thus, they nurture the system with relevant data. From the stakeholders previously identified, these are the national authorities (which we refer to as official data producers) and researchers, NGOs or any other unofficial source (which we refer to as unofficial data producers). One key aspect of the project is to identify how to manage (i.e., gather, homogenize and store) such data.
- Data consumers that will analyze the available information according to their granted access to data. We further classify the data managed by the system according to access grants: open data (i.e., available to anyone), research-related data and sensitive data. The first kind is typically accessed by regular users (e.g., concerned citizens), who want to access pieces of public information. The main indicators / statistics about Chagas will be open for everyone. For the second kind, we refer to them as data analysts who aim at performing advanced analytical queries and gain insight into data (typically, researchers and NGOs). Thus, an authorised account will be needed to access this data. For the third kind, this data is only available for the official data producer (e.g., a national health ministry) and the WHO technical groups.

Related to this data classification, although not stakeholders, we also identify the following relevant actors (further details in Sect. 5.1):

- Data administrators who will take care of consistency of data inserted in the system (e.g., by applying protocols, granting certifications, etc.). These role is exclusively played by a specific technical group (IT-oriented).
- The system administrator, who is an IT person, that will manage and maintain the system from the ICT point of view.

– Other Information Systems: quite popular nowadays and to be used to also provide relevant information such as ProMed, PubMed and Google Alert information systems.

A detailed description of the data workflows and the interaction of the stakeholders with the system can be found in Sect. 5.3.

3.3 CID Project IT Facet

We analysed information systems and technologies used in WHO that would help us understand the requirements and challenges involved in the development of the system.

Global Health Observatory. The Global Health Observatory (GHO) is part of the World Health Organization (WHO). GHO is the WHO's gateway to health-related statistics from around the world. Therefore, the main goal of GHO is to provide a easy access to:

– Country data and statistics with focus on comparable estimates.
– WHO's analyses to monitor global, regional and country situation and trends.

GHO presents itself in four different sections, GHO theme pages, GHO database, GHO country data and GHO issues analytics reports. GHO theme pages covers global health priorities such as the health-related Millennium Development Goals, mortality and burden of disease, health systems, environmental health, noncommunicable diseases, infectious diseases, health equity and violence and injuries. GHO theme pages present:

– Highlights showing the global situation and trends, using regularly updated core indicators.
– Data views customized for each theme, including country profiles and a map gallery.
– Publications relevant to the theme.
– Links to relevant web pages within WHO and elsewhere.

The GHO database provides access to an interactive repository of health statistics. Users are able to display data for selected indicators, health topics, countries and regions, and download the customized tables in diferent formats such as Excel, CSV, HTML and GHO XML. The GHO country data includes all country statistics and health profiles that are available within WHO. The GHO issues analytical reports on priority health issues, including the World Health Statistics annual publication, which compiles statistics for key health indicators. Analytical reports will address cross-cutting topics such as women and health.

Standards. The WHO Indicator and Measurement Registry (IMR) is a central source of metadata of health-related indicators used by WHO and other organizations. It includes indicator definitions, data sources, methods of estimation and other information that allow users to get better understanding of their indicators of interest.

It facilitates complete and well-structured indicator metadata, harmonization and management of indicator definitions and code lists, internet access to indicator definitions, and consistency with other statistical domains.

It promotes interoperability through the SDMX-HD indicator exchange format and allows incorporation of appropriate international standards such as SDMX MCV (Metadata Common Vocabulary), ISO 11179 (Metadata Registry), DDI (Data Documentation Initiative) and DCMES (Dublin Core).

The SDMX-HD is a data enchange format (Statistical Data and Enchange Metadata) based on the data enchange format SDMX. SDMX-HD intends to serve the needs of the Monitoring and Evaluation community. It has been developed by WHO and partners to facilitate the enchange of indicators definitions and data in aggregate data systems.

The SDMX standard describes statistical data and metadata through a data structure definition (DSD), which defines concepts that define dimensions, attributes, codelists and other artefacts necessary to describe the structure and meaning of data. A parallel metadata structure definition (MSD) describes metadata associated with data at observation, series, group, and dataset levels.

Data Repository. The GHO data repository provides access to over 50 data-breaksets, which can be selected by theme or through a multi-dimension query functionality. The datasets contains information about mortality and burden of diseases, the Millennium Development Goals (child nutrition, child health, maternal and reproductive health, immunization, HIV/AIDS, tuberculosis, malaria, neglected diseases, water and sanitation), non communicable diseases and risk factors, epidemic-prone diseases, health systems, environmental health, violence and injuries, equity among others. In addition, the GHO provides online access to WHO's annual summary of health-related data for its 194 Member states: the World Health Statistics 2012.

GHO has only information about eight of the seventeen neglected tropical diseases, as follows: Buruli Ulcer, Dracunculiasis, Human African Trypanosomiasis, Leprosy, Lymphatic Filariasis, Schistosomiasis, Soil-transmitted helminthiasis, and Trachoma.

The information provided by GHO can vary. In some cases GHO provide information about the number of new reported cases for a disease (e.g., Buruli Ulcer) and in other cases can provide country data. Therefore, the information that can be found in the NTDs section is:

- Country data.
- Status of endemicity.
- Annual incidence of cases by country.
- Number of reported cases.
- Population living in endemic areas.
- Population treated.

Country Data. The Country data table shows countries in rows and indicators in columns for a particular year. The indicators may vary between diseases. Figure 5 shows an example of country data for schistosomiasis.

	Country	Year	SAC population requiring PC for SCH annually	Population requiring PC for SCH annually	Number of people targeted	Reported number of people treated	Age group	Reported number of SAC treated	Programme coverage	National coverage
AFR	Angola	2006	-	-	-	-	-	-	-	-
AFR	Angola	2007	-	-	-	-	-	-	-	-
AFR	Angola	2008	-	-	-	-	-	-	-	-
AFR	Angola	2009	-	-	-	-	-	-	-	-
AFR	Angola	2010	2,620,044	4,722,157	-	-	-	-	-	-
AFR	Angola	2011	2,680,353	4,849,215	-	-	-	-	-	-
AFR	Benin	2006	-	-	-	-	-	-	-	-
AFR	Benin	2007	-	-	-	-	-	-	-	-
AFR	Benin	2008	-	-	-	51,433	SAC	51,433	-	0.69%
AFR	Benin	2009	-	-	-	-	-	-	-	-
AFR	Benin	2010	1,211,805	2,263,785	400,475	364,697	SAC	364,697	91.07%	16.11%
AFR	Benin	2011	1,246,286	2,333,298	-	-	-	-	-	-
AFR	Bctswana	2006	-	-	-	-	-	-	-	-

Fig. 5. Country data (Schistosomiasis disease) [10]

Status of Endemicity. Status of endemicity table shows countries in the rows and year in columns. The value for a concrete year and country could be endemic or non-endemic. It is possible to filter the data by country, year, WHO region and World Bank income group. This table can be downloaded in different formats: CVS, Excel, HTML, and GHO-XML. Figure 6 shows the status of endemicity for Trachoma disease.

Number of New Reported Cases. The number of new reported cases table shows countries in rows and years in columns. The value for a concrete country and year can be a natural number or Data not reported. It is possible to filter the data by country, year, WHO region and World Bank income group. Also, it is possible to download de data in different formats, CSV, Excel, HTML and GHO-XML. Figure 7 shows the number of new reported cases for Buruli Ulcer.

Population Living in Endemic Areas. The population living in endemic areas table shows countries in rows and years in columns. The value for a concrete country and year can be a natural number or No data available. It is possible to filter the data by country, year, WHO region and World Bank income group. Also, it is possible to download de data in different formats, CSV, Excel, HTML and GHO-XML. Figure 8 shows the population living in endemic areas table for Trachoma disease.

Country	Status of endemicity for blinding trachoma[	
	2010
Slovenia	Non-endemic
Solomon Islands	Endemic
Somalia	Endemic
South Africa	Non-endemic
Spain	Non-endemic
Sri Lanka	Non-endemic
Sudan	Endemic
Suriname	Non-endemic
Swaziland	Non-endemic
Sweden	Non-endemic
Switzerland	Non-endemic
Syrian Arab Republic	Non-endemic

Fig. 6. Status of endemicity (Trachoma disease) [10]

Country	2008	2009	2010	2011	2012
Equatorial Guinea	rted	Data not reported	Data not reported	Data not reported	0	Data not reported
French Guiana	2	8	2	7	3	2
Gabon	32	53	41	65	59	45
Ghana	668	986	853	1048	971	632
Guinea	rted	80	61	24	59	82
Indonesia	rted	Data not reported	Data not reported	Data not reported	Data not reported	Data not reported
Japan	3	2	5	8	10	4
Kenya	rted	Data not reported	Data not reported	Data not reported	Data not reported	Data not reported

Fig. 7. Number of new reported cases (Buruli Ulcer) [10]

Country	Population living in trachoma endemic areas[				
	2006	2007	2008	2009	2010
Brazil	No data available	27000000	27000000	27000000	27000000
Burkina Faso	No data available	3978611	5299721	5299721	7308338
Burundi	No data available	No data available	1871208	1871208	941085
Cambodia	2125447	2125447	2125447	2125447	2416543
Cameroon	No data available	No data available	No data available	5276973	No data available

Fig. 8. Population living in endemic areas (Trachoma disease) [10]

Population Treated. The population treated table shows countries in rows and years in columns. The value for a concrete country and year can be a natural number or No data available. It is possible to filter the data by country, year, WHO region and World Bank income group. Also, it is possible to download de data in different formats, CSV, Excel, HTML and GHO-XML. Figure 9 shows the population living in endemic areas table for Trachoma disease.

	Population treated for active trachoma[*]				
Country	**2005**	**2006**	**2007**	**2008**	**2009**
Djibouti	No data available	No data available	No data available	No data available	72
Egypt	No data available	No data available	No data available	No data available	No data available
Eritrea	No data available	No data available	1364817	1364830	0
Ethiopia	2618488	2600000	6044714	14832830	15695222
Fiji	No data available	No data available	No data available	No data available	No data available
Gambia	18402	401397	143658	151135	111969
Ghana	796378	770124	899580	147122	Case management
Guatemala	No data available	No data available	No data available	No data available	No data available
Guinea	0	0	0	0	0

Fig. 9. Population treated (Trachoma disease) [10]

Map Gallery. The GHO map gallery includes an extensive list of major health topics. Maps are classified by the following themes: Alcohol and Health, Child Health, Cholera, Environmental health, Global influenza virological surveillance, Health systems financing, HIV/AIDS, Malaria, Maternal and reproductive health, Meningococcal meningitis, Mortality and global burden of disease (GBD), Neglected tropical diseases (NTDs), Noncommunicable diseases, Road safety, Tobacco control and Tuberculosis[6].

The GHO map gallery allows searching maps by geographical region, topics and keywords. Figure 10 shows a possible search defining region as world, topic as neglected tropical disease and keyword as Chagas.

The neglected tropical disease map gallery contains maps about NTDs topics such as distribution of a disease in a concrete year (Fig. 11), new cases reported (Fig. 14), facilities involved (Fig. 13) and prevalence rate (Fig. 12).

Country Statistics. The country statistical pages bring together the main health data and statistics for each country, as compiled by WHO and partners in close consultation with Member States, and include descriptive and analytical summaries of health indicators for major health topics. Each country statistics page has three main sections:

---

[6] http://gamapserver.who.int/mapLibrary/

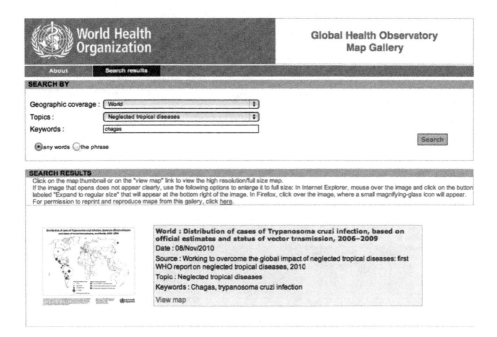

Fig. 10. GHO map gallery search engine [10]

Distribution of Buruli ulcer, worldwide, 2011

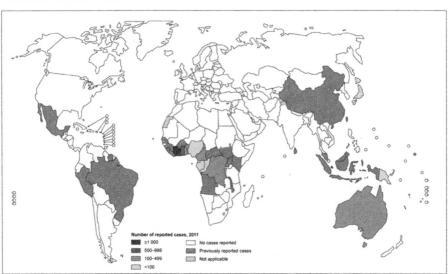

Fig. 11. Distribution of Buluri Ulcer [10]

Facilities involved in Buruli ulcer clinical trial 2012-2015

Fig. 12. Facilities involved in Buruli Ulcer [10]

Leprosy prevalence rates, data reported to WHO as of January 2012

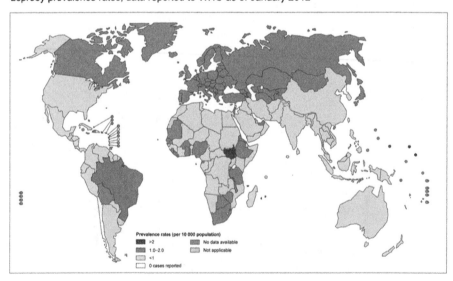

Fig. 13. Leprosy prevalence rates [10]

Leprosy new case detection rates, data reported to WHO as of beginning January 2011

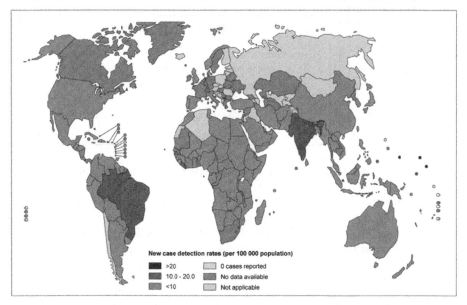

Fig. 14. Leprosy new case detection rate [10]

1. Country data and statistics
2. Country profiles
3. Primary metadata

3.4 CID Project Development Facet

Due to the complexity of the project, at this stage it was decided to estimate only the costs involved for the requirements part of the project and to postpone the estimation of the cost of the whole project until the end of the requirements engineering process.

4 Core Activities

This block of Pohl's framework is composed of the three main core activities, namely, Documentation, Elicitation and Negotiation, as shown in Fig. 15.

These are described below:

1. **Elicitation.** The goal of this activity is to elicit the requirements at all the levels of the system. These requirements may exist in different forms, for instance: in ideas, intentions, requirements models, in existing systems, etc. In addition, there are many sources (stakeholders, documents, existing systems) that have to be elicited in order to gather all the possible requirements to support the acceptance of the system.

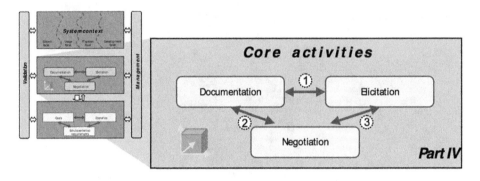

Fig. 15. Core activities in Pohl's requirements engineering framework [2]

2. **Negotiation.** Each stakeholder has different wishes and needs that may be in conflict with each other; reason why negotiation aims to achieve an agreement among them. Conflicts must be resolved involving the most relevant stakeholders while also these may be seen as an opportunity to create also new ideas.

3. **Documentation.** The main aim of this activity is to document the elicited requirements according to the rules. According to Klaus Pohl, documentation is important in order to establish a common reference, promote communication between stakeholders, support training for new employees and to preserve expert knowledge. Rules for documentation may be either:
   - General Rules (e.g., define layout, headers, version history),
   - Documentation Rules (e.g., rules to ensure quality of template for the use of elicitation and negotiation), or
   - Specification Rules (more restrictive than documentation rules, e.g., rules for subsequent development activities like the use of requirements specification language).

   In order to avoid lexical ambiguity, the author suggests glossaries may be used along with synonyms and related terms.

In general, these activities were not much different of any traditional elicitation, negotiation and documentation activities. However, there are two main differences that must be considered when dealing with DSS. On the one hand, the heterogeneity of the stakeholders involved. In general, in DSS, and specifically, in the CID project, the point of view of each stakeholder is drastically biased by previous experiences and tend to simplify other factors that are not releveant to him / her. In our case, we realised that not even the WHO had a clear view of all the factors affecting Chagas. This difficulty of understanding each other was not only a challenge for IT people, but also for other actors involved. For example, entomologists and doctors (among whom we must distinguish specializations) also had problems to combine their visions. Due to this complexity we identified the most relevant techniques and methods to elicit and negotiate in DSS projects (see Sects. 4.1 and 4.2). On the other hand, the real difference comes from the artifacts produced during these activities (see Sect. 5.3 for further details).

4.1   CID Project Elicitation

Once described the system context of the project, software analysts with the help of health and vectorial experts trawled the work to learn and understand the requirements of the project, i.e. elicit the requirements of the system.

In addition to brainstorming the elicitation techniques used were the following:

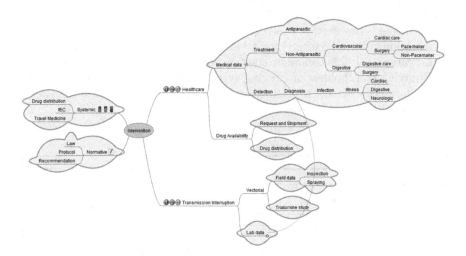

Fig. 16. Mind map produced to identify the main factors affecting Chagas in the intervention package (result of one brainstorming session)

1. Interviews [11]: software analyst performed several interviews to health and vectorial experts. Most of them were exploratory interviews, i.e. conversations by means of which the interviewer elicited information about the opinion or view of the interviewee with respect to some issue. The results of such interviews were qualitative. For example, software analyst interviewed a health expert in co-infection to understand all the variables to take into account when defining the part of the conceptual model related to diagnosis.
2. Workshops [12]: around 4–5 people joined during several 2–3 days sessions to refine the requirements. Some of the typical different assistance techniques used were brainstormings, iterative goal and scenario definitions, discussions and work in sub-groups and presentation of the results in the plenum. An example of the agenda for one of the workshops was:

9:00 - 9:30 Welcome session and presentation of the goals of the workshop: information required about Diagnosis, Drug Availability, Vectorial, Treatment, Systemic and Normative Intervention.
9:30 - 10:00 Brainstorming of the goals of the workshop
10:00 - 10:45 Categorisation of the brainstorming results,
10:45 - 13:00 Grouping and refinement of the goals, assignment of goals to subgroups,

13:00 - 14:30 Lunch break

14:30 - 18:00 Work in sub-groups: scenario definition for the elicitated goals

18:00 - 19:00 Presentation of the intermediate results by each sub-group.

3. Questionnaires [11]: questionnaires were suitable when physical interviews or workshops were not feasible. Basically, lists of questions from the software analyst to the health or vectorial experts were sent by email. Initially, software analysts tried to ask specific questions related to the conceptual model being defined. However, it was required to change to more user friendly documentation (basically, forms) for a better understanding of the questions.

4. Observation [13]: this technique consists on an observer eliciting requirements by studying stakeholders or existing systems. Even though there is no a current information system to eradicate Chagas, similar projects have been developed for other diseases within the WHO context.

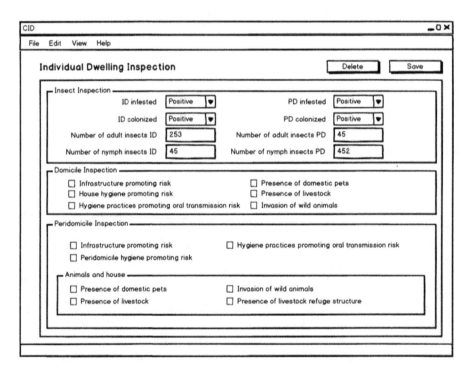

Fig. 17. Mockup of the collection of information for Individual Dwelling Inspection

The elicitation techniques described above were also supported by the following assistance techniques [2]:

1. Brainstorming [14]: Brainstorming was basically performed during the first workshops for defining the main goals of the project and clarification of ideas and concepts.

2. Mind mapping [15]: We used the FreeMind tool[7] to represent the context of the project and to group the different use cases that were needed to develop in the whole project as shown in Fig. 16.
3. Prototyping [16]: this technique was completely necessary to support the definition of the use cases and the conceptual model of the project. Conceptual models in UML were difficult to understand for Vectorial and Health experts, but the creation of mockups (with the Balsamiq Mockup tool)[8] facilitated the feedback from them. We created them for each subpart of the conceptual model. An example is shown in Fig. 17.

4.2   CID Project Negotiation

In this project, the main conflict was to reach a consensus between the value obtained to collect as much information as possible to the maximum level of detail from the countries to the actual ability to collect such information. For example, at the very beginning the health expert asked that the system collected details of each dwelling inspection on a vectorial control. However, to ask for so much information and at this level of detail entailed the risk that the system would never be used or that the information entered would not be accurate enough. The use of mockups, as mentioned above, were indispensable to resolve these conflicts and reach consensus. Finally, priority was given to get the minimum information necessary to obtain the maximum impact to combat Chagas.

This conclusion is extensible to any DSS. In any DSS the trade-off of value added versus difficulty to gather the needed data must be analyzed in depth and negotiated. At the negotiation stage, it must clearly be defined what is the value added and weight it against the difficulty to obtain such information. For example, ideally, some stakeholders were asking for several parameters at house level when gathering information about infestation. However, after showing some mockups of the forms the field teams should fill it was clear for all of us that it was unfeasible and, in practice, most teams would refuse to fill all that information per house (due to its time-consuming nature). It was more advisable to gather data at a certain aggregation level, with few (but highly relevant) parameters. In this precise example, this was negotiated and agreed by presenting a mockup (that of Fig. 17) and modify it until reaching consensus.

4.3   CID Project Documentation

Several documents were generated during the whole project: for describing the system context, the goals, the scenarios and the specification of the solution-oriented requirements. The latter were documented following the IEEE 830–1998 software requirements specification standard.

---

[7] http://freemind.sourceforge.net/.

[8] http://www.balsamiq.com.

Table of Contents

The requirements were defined at a level of detail that permits designing and implementing the software system that meets the defined requirements. It includes the definition of the function systems, the conceptual model, and the software design of the system.

5   Requirement Artifacts

The term "requirements artifacts" stands for a documented requirement using a specific format. In this section we differentiate between three different requirements artifacts according to Klaus Pohl, namely goals, scenarios, and solution-oriented requirements [2] and shown in Fig. 18:

- **Goals:**
  In the requirements engineering domain, the stakeholder's intentions about the objectives are documented as goals stating what exactly is expected or required from the system.

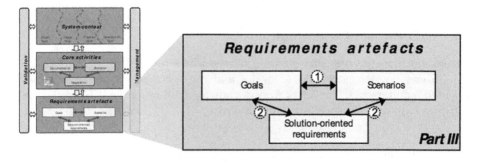

Fig. 18. Requirements artifacts in Poh'ls requirements engineering framework [2]

– **Scenarios:**
  As mentioned above, goals are used to document the stakeholder's intentions. Therefore, scenarios are further used to document intentions by illustrating if the scenario satisfies or fails to satisfy the goal by representing the interactions between the system and its actors.

– **Solution-Oriented Requirements:**
  As mentioned above, goals and scenarios are the basic foundations for developing solution-oriented requirements. In contrast to goals and scenarios, solution-oriented requirements are created to specify a deeper required level of details. By joining these three together requirements, we may define the reference software engineers to implement the system.

5.1 CID Project Goals

To define the goals of each actor we classified the actors that would interact to the system in three types of users and three types of systems as shown in Fig. 19.

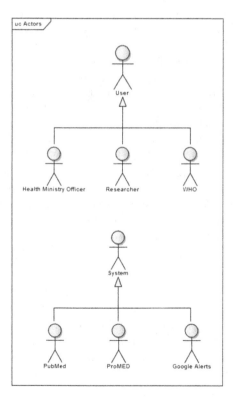

Fig. 19. Actors of the CID system

For each one, we defined in natural language his or her goals.

The main goals of the WHO actor are to visualise, manage, analyse and import/export information of the system. On the one hand, since WHO must have access to the entire information system database, he or she should be able to consult, but not modify, any type of information related to any country. Additionally, WHO actor will be able to manage assessments, regulations and certifications. On the other hand, he or she can generate and visualise maps and diagrams with any type of information gathered in the system. Finally, he or she will be able to import and export any type of information provided by countries or researchers by uploading it with an Excel file.

The HMO actor, in a similar way as the WHO actor, should interact with the system in order to visualise, manage and import/export information. Health Ministry Officers should be allowed to provide, modify, and delete information related to his own country, therefore he or she will not be able to consult, modify and delete information from other countries. Information he or she can manage ranges from healthcare, through transmission interruption, to systemic and normative. In particular, he or she can manage the following information: treatments, diagnosis, insecticide applications, triatomine bug studies, inspections, regulation and systemic information. In addition, he will be able to generate and visualise maps and diagrams with the information he provided to the system. Finally, he will be able to import and export any type of information by uploading it with an Excel file.

The Researcher actor, in a similar way as the other actors, interacts with the system in order to visualise, manage, import/export, and analyse information. Researchers can provide information related to healthcare, and transmission interruption. In particular, they can manage the following information: treatments, diagnosis, insecticide applications, triatomine bug studies and inspections. In addition, they will be able to generate and visualise maps and diagrams with the information they provided to the system. Finally, they will be able to import and export any type of information by uploading it with an Excel file.

Actors may represent roles played by human users, external hardware, or other subjects such as information systems. The ProMED, PubMed and Google Alert information systems are also considered actors of the system. All these three actors interact with the system notifying it when new cases of Chagas disease appear.

5.2   CID Project Scenarios

Once defined the main goals of the actors, the next step was to look for the system's scenarios by means of use cases. We needed several iterations with the WHO experts before we arrived to the final use cases version. Furthermore, all the use cases defined were classified in four groups: management of tables, visualization of information, exportation and importation of information, and analysis of information.

The following list presents all uses cases for each of the previous groups:

- **Management of Tables**: Manage Regulations, Manage Assessments, Manage Certifications, Manage Triatomine Study, Manage, Dwelling Triatomine

Study, Manage Collective Triatomine Study, Manage Insecticide Application, Manage Collective Insecticide Application, Manage Dwelling Insecticide Application, Manage Inspections, Manage Dwelling Inspections, Manage Collective Inspections, Manage Diagnosis, Manage Individual Diagnosis, Manage Collective Diagnosis, Manage Collective Oral Diagnosis, Manage Treatments, Manage, Individual Treatments, Manage Collective Treatments, Manage Drug Distribution, Manage Systemic Information, Manage Associations, Manage IEC, Manage Research, Manage Drug Distribution System, Manage Lab Quality Control and Manage Health Economic.

– **Visualization of Information**: Visualise Diagrams and Visualise Maps.
– **Import/Export Information**: Export Information and Import Information.
– **Analysis of Information**: Analyse Data

We show next an example of the Use Cases defined:

Use Case Name
Manage Individual Diagnosis
Active Actor
Researcher
Trigger
Researcher indicates manage individual diagnosis
Preconditions
The Researcher must be identified and authenticated.
Stakeholders and interests

1. Researcher: manage (create, modify or delete) individual diagnosis information he provided.

Main Success Scenario

1. User indicates to manage individual diagnosis
2. System presents all registers of available individual diagnosis gathered into the system
3. User selects an existing individual diagnosis register
4. System presents the individual diagnosis information
5. User updates information about the individual diagnosis and then indicates save the individual diagnosis register
6. System updates the register

Extensions

3a. Create a new individual diagnosis register.

1. User indicates to create a new individual diagnosis register
2. System presents a form in order to fill in with the individual diagnosis information
3. User enters information about the individual diagnosis and confirms the creation of the individual diagnosis register

Return to step 6 of Main Success Scenario

5a. Delete individual diagnosis register

1. User indicates to delete the individual diagnosis register
2. System asks for confirmation
3. User confirms deletion of the individual diagnosis register

3a. No confirmation

1. User cancels the action

Return to step 4 of Main Success Scenario
Return to step 6 of Main Success Scenario
Outcome
An individual diagnosis register has been created, updated or deleted with the information provided by user.

5.3   CID Project Solution-Oriented Requirements

Solution-oriented requirements define the data perspective and the functional perspective on the system. The data perspective focuses on defining the data/ information to be managed by the system. We defined this information, the conceptual model, by means of UML class diagrams [17]. The aim of said conceptual model was to define and represent real-world concepts, that means, define relationships between different concepts of Chagas disease and WHO, and describe the meaning of terms and concepts used by health specialists about Chagas disease. Additionally, the conceptual model of the information system was structured in different packages to make easier the comprehension of all its concepts. Therefore, the following packages were defined: Healthcare package, Transmission Interruption package, Systemic package, Normative package, Geographic and Temporal package, Wildlife package and the Bibliographic package.

An example of a fragment of the conceptual model is shown in Fig. 20.

The functions of the systems were described by refining the UML use case diagrams: for the most relevant functions of the system, sequence diagram with operation contracts in OCL [18] were specified.

Regarding the design, the system functional architecture is described in Fig. 21 and it is divided in 4 main layers, namely:

- The data input layer through which relevant data related to the Chagas disease is introduced in the system,
- The integration layer (or data warehouse), which shows a single, unified view of the data gathered,
- The administration layer, which monitors the system to detect relevant events, and
- the exploitation layer, which is meant to query and perform advanced analytics over the gathered data.

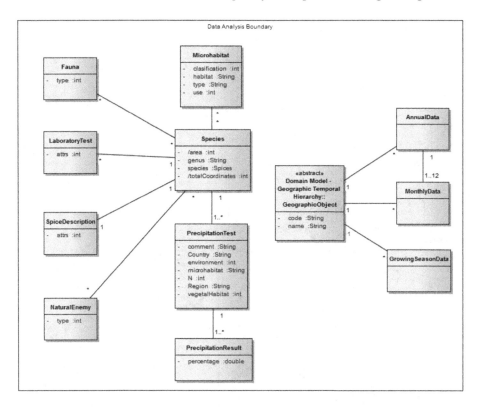

Fig. 20. Wildlife package's conceptual model

This architecture also refines the use cases discussed above. The data input layer is meant to let end-users introduce their data into the system. Data gathered is properly identified by the person / group / organization uploading such data and arranged in 5 different subsystems within the data warehouse: healthcare, transmission interruption, wildlife, normative and systemic.

- Healthcare: This subsystem contains information about healthcare indicators, and is divided in three categories: patients diagnosis and treatment, and drug delivery. For the first two, data is gathered either at the individual level (i.e., data is related to an anonymized person and indicators such as age, gender, nationality or phase of the patient, among others, are stored) or collective level (i.e., data related to a geographic area and aggregated measures such as number of infected, coinfected or evaluated people, among others, are stored). For the latter, the system stores data related to medicines and their distribution.
- Transmission interruption: Although Chagas disease can be transmitted by different routes, this subsystem focus on the vectorial transmission and contains information related to dwelling inspections, dwelling insecticide applications and triatomine bug studies. The three of them can be reported either at the individual or collective level. Individual data reported refers to a specific dwelling and indicators such as the number of adult kiss bugs and nymphs

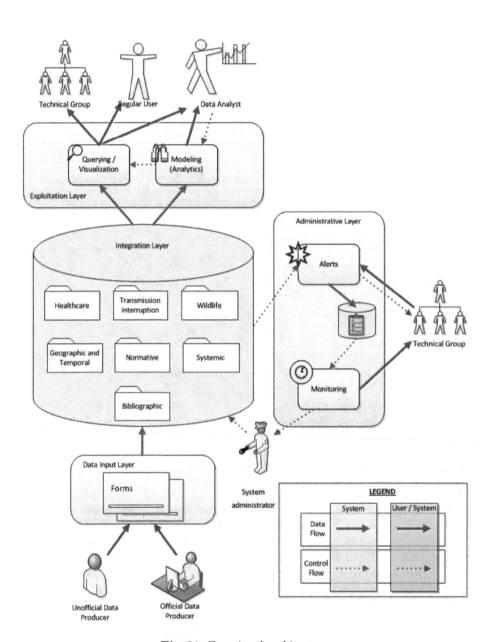

Fig. 21. Functional architecture

found or bugs infested with the *T. cruzi* parasite, among other indicators, are stored. Collective data is related to a geographic area and aggregated measures such as number of inspected, infested, or colonized dwellings, among others, are stored).

– Wildlife: This subsystem stores data about flora, fauna and climatology. In order to foresee to what regions the Kiss Bug could spread in the near future information about all known species which can transmit the Chagas disease as well as climatological factors are stored.

– Normative: This system stores data about regulations, certifications and assessments. Regulations are typically provided by official users, whereas certifications and assessments are introduced by WHO.

– Systemic: Data about the activities and procedures that countries are implementing in order to control and eliminate the Chagas disease. This includes data such as recognized associations, research capacity, data on IEC (Information, Education, Communication), etc.

Relevantly, any piece of data introduced in the system is always characterized with a geographical and a temporal object, which are managed by the geographical and temporal subsystem. Finally, the system is prepared to allow researchers introduce complementary data to the official data uploaded by countries. Research data can be backed up with external publications (e.g., journal papers) and for these cases the bibliographic subsystem is responsible for storing and managing the bibliography. Any uploaded research data is automatically put in quarantine until validated by the corresponding technical group who can decide to either make it available, once manually checked its quality, or kept in quarantine if its veracity is doubtful. This validation task is performed by means of the monitoring subsystem available under the administration layer, which is also responsible to support predefined monitoring alerts. In short, the technical group may decide to set up some high-level protocols. For this milestone we implement ECA (event-condition-action) rules. In presence of a certain event (e.g., a new infection case is inserted), if a certain condition is given (e.g., this infection case is associated to a country -geographical object-, which had no previous infection cases), an automatic alert is triggered (e.g., an email is sent) to the technical group. The technical groups is responsible for defining which events and conditions must be monitored through the alerts module, which, in turn, will store the ECA rules in the metadata repository. The monitoring module monitors the right piece of information from the data warehouse and an alert is triggered if the ECA rule holds. Monitored data can be consulted by the technical group at any time.

Last but not least, the exploitation layer is responsible for querying / visualizing and performing advanced analytical tasks on the data; the ultimate goal of the whole system. According to the granted access rights of each kind of user a user may query and visualize that portion of data available to him / her. Then, the same module is responsible to render the query result. Alternatively, data analysts need advanced tools, such as R, to create models of different aspects related to Chagas in order to foresee the impact of certain actions / inactions, whose results are eventually rendered by means of the querying / visualization module.

6   Validation and Management

Validation aims to state if the input and output (requirements artifacts) of the core activities satisfy the criteria established. This phase of validation is done by involving different resources, such as: stakeholders, requirement sources (laws and standards) and external reviewers. Validation answers the question: Am I building the right system?

Apart from Validation, Management is another one of the cross-sectional activities in the requirements engineering framework.

6.1   CID Project Validation and Management

In this project, following principles of requirements validation according to Pohl, we ensured that relevant and appropriate WHO Health, Vectorial and software experts were involved for each aspect of a requirement artefact to be checked during validation; we considered changing the documentation format of the requirements into a format that matched the validation goal and the preferences of stakeholders who performed the validation and; we also established guidelines that clearly determined when or under what conditions an already released requirements artefacts had to be validated again. We mainly used techniques of inspection, desk-check, walkthrough and prototyping when appropriate.

The goal of management in requirements engineering is to (1) observe the system context to detect context changes, (2) manage the execution of requirements engineering activities and (3) manage the requirement artefacts. The first goal has not been a big deal, since as part of their own work, health and vectorial experts have been participating in the worldwide events related to any possible law, policies or medical changes that could affect the project and software specialist have been analysing the emergent technologies that could be used in the implementation of the project. The most difficult task in the management of the execution of requirements engineering activities has been to make-up the agendas and trips of the stakeholders for the workshops. Finally, for the management of the requirement artefacts, we used, on the one hand, TortoiseSVN[9], a free open-source Windows tool for the Apache Subversion control system for controlling the different versions of all documentation created and, on the other hand, to ensure the traceability among the different software models, all of them were defined in the Enterprise Architect, a collaborative modeling, design, and management platform based on UML 2.4.1 and related standards from Sparx Systems[10].

---

[9] http://tortoisesvn.net/.
[10] http://www.sparxsystems.com/.

7   Conclusions and Further Research

The CID project was useful as a case study of the application of Pohl's framework to obtain the requirements of a decision-support system. The complexity found in the CID project was an example of what is expected to deal in most of these projects:

- **Complexity of decision-support systems**: Unlike operational systems, decision-support systems such as data warehousing are a complex and expensive component of the corporate's IT infrastructure involving a conglomerate of extraction, cleansing and transforming systems, Data Warehouse database(s), Data Marts, metadata management system, usually presented in a summarized format for visualization purposes [19]. For instance, the CID project consisted of gathering information from heterogeneous sources such as: hospitals using databases to record the infected patients, researchers using excel forms to record data about the raids and dwellings inspections, and health minister officers managing information about regulations in PDF format; making this a complex process to build the CID system.
- **Integration with diverse data sources**: Decision-support systems collect data from heterogeneous sources including the operational systems or other external sources. Therefore, procedures and requirements must be specified for the data collection and integration [20]. Plus interfaces and connections to already existing systems (inside or outside the organization) must be considered [19]. During the CID project, heterogeneous sources mentioned in the previous example had to be integrated to present a single unified interface; thus, problems of heterogeneous source integration were present such as: *technical heterogeneity* (integrating different file formats and databases such as Microsoft Excel, PDFs, and SQL Server databases) and *semantic heterogeneity* (identifying and integrating concepts from the sources which semantically correspond to one another, such as "kiss bug" in the U.S.A is semantically equivalent to "vinchuca" in Argentina).
- **Expressing requirements of decision-support systems**: At early stages of the decision-support systems project, the future users of the system have difficulties to express their requirements [21]. In contrast to operational systems which are based on consistent specifications [19]. For instance, in the CID project, the stakeholders aimed to collect as much information as possible with the maximum level of detail (e.g., number of insects per house). However, it is too ambitious to obtain this level of detailed data, plus the system could become too complex. Hence, an agreement among the stakeholders had to be reached by negotiating the right balance of having information with enough level of detail (i.e. aggregated data per certain area) to impact the main objective of elimination of Chagas. As believed by Paul Valéry: *"What is simple is wrong, what is complicated is useless"* [22].
- **Identifying relevant subjects and perspectives of analysis**: The multidimensional model has become the de facto standard for decision-support systems that categorize the data as facts (representing a business activity) or

dimensions (perspective of the data) being usually textual [23]. While analyzing user requirements, in decision-support systems it is crucial to identify the subject of analysis by perceiving the metrics behind the user demands. Also to determine the extent to which a metric can be operated (summarized, counted, etc.) along with the different perspectives from where to analyze the relevance subjects of analysis [20]. As discussed later, this specific analysis-oriented conception requires considering specific data models.

- **Data exploration and summarization**: It is needed to firstly discuss the relevance of exploring data from different points of view, meaning changing the data granularity (data detail level) on demand. To do so, summarizability (assuring the correctness of summary operations) is a must; specifically to validate the correctness of aggregated results for whichever combination between facts and dimensions [20]. Summarizability is done in order to avoid erroneous conclusions and decisions [24]. Thus, it is important to express the requirements constraint regarding the aggregated data. For instance, in the CID project, visual representations of hierarchical structures in space and time were required, for instance, showing "Infected Dwellings" for Argentina regions over a period 2008–2014 to detect potential infected geographical areas.

- **Fast track of user requirements changes**: Decision-support systems are in constant evolution and must face changes during its development that may impact on the models built to meet the end-user requirements [20]. The issue is that the relationship between the requirements, elements in the model and data sources are lost in the process since no traceability is specified neither documented [25]. Accordingly, track of all the requirements and changes must be kept in decision-support systems during the complete project's life-cycle. For instance, showing requirements dependencies with cross-references analysis to demonstrate the impact of a certain change in a fact or dimension (e.g., traceability matrices) [26].

- **High-quality documentation and standard vocabulary:** Decision-support systems projects use preexisting operational documentation in order to define the data extraction and integration procedures; however, legacy documentation usually lacks quality, which makes the task of extracting the requirements harder [20]. In addition, a vocabulary, is necessary to identify definitions, acronyms and abbreviations for establishing lexis to be shared among all parties related to the project [27]. For instance, during the CID project, health specialists (e.g., entomologists) certainly use medical and domain terminology for their every day communication such as "trypanosomiasis and Triatominae", which software analysts and neither doctors were able to understand, and neither to match these with the heterogeneous sources. For this reason, there is a huge need to bridge the gap between stakeholders and IT people by creating a Vocabulary. However, not much advantage can be taken from the vocabulary in a plain document, but certainly there is a need to translate this vocabulary into a machine-readable format with a cross-reference structure to be used when creating the future schema for the system.

The project clearly motivated to do further research about the current approaches of requirements engineering in decision-support systems: we needed to review the literature on the topic to obtain a good grasp of the main published work. In short, many different aspects of the whole requirement engineering cycle have been presented but there is no overall and comprehensive framework such as Pohl's framework for decision-support systems. After carefully reading the literature related to the requirements engineering framework, we noticed that we would achieve a more understandable classification by organizing the literature we found according to the building blocks proposed in the framework. We believed that the use of this framework would be a good fit for our study due to the following reasons:

1. As mentioned previously, the first and main reason is that the framework defines the major structural building-blocks and elements of the requirements engineering process (e.g., elicitation, negotiation, documentation, goals, etc.).
2. It provides a well-structured base for the fundamentals, principles and techniques of requirements engineering.
3. It is not adhered to a specific methodology or a type of software project.
4. In addition, this framework consolidates various research results and has been proven to be successful in a number of organizations for structuring their requirements engineering process.

For the aforementioned reasons, the use of this framework would provide a better organization and structure of the literature found (requirements engineering for decision-support systems).

As a natural continuation of this work, and after all the lessons learnt, we are formalizing our findings in the form of a comprehensive framework, following the lines above describe, within the context of an IT4BI Master Thesis of one of the authors of this paper (S. García) and supervised by other two (O. Romero and R. Raventós). The expected outcome will be a systematic generic approach entitled "Requirements Engineering for Decision-Support Systems" (RE4DSS) based on the thorough view of the domain gained during the analysis of the literature.

References

1. World Health Organization: Working to overcome the global impact of neglected tropical diseases: First WHO report on neglected tropical diseases. Geneva, Switzerland (2010)
2. Pohl, K.: Requirements Engineering - Fundamentals, Principles, and Techniques. Springer, Heidelberg (2010)
3. Aurum, A., Wohlin, C. (Eds.): Engineering and Managing Software Requirements. Springer, Heidelberg (2005)
4. IEEE.: IEEE Std 610.12-1990(R2002). IEEE Standard Glossary of Software Engineering Terminology (1990)
5. Coura, J.R., Viñas, P.A.: Chagas disease: a new worldwide challenge. Nature **465**, S6–S7 (2010). (n7301_supp)
6. Peters, W., Gilles, H.M., et al.: A Colour Atlas of Tropical Medicine and Parasitology. Wolfe Medical Publications Ltd, London (1977)

7. Feasey, N., Wansbrough-Jones, M., Mabey, D.C., Solomon, A.W.: Neglected tropical diseases. Br. Med. Bull. **93**(1), 179–200 (2010)
8. Ferreira, M.S., Nishioka, S.A., Silvestre, M.T.A., Borges, A.S., Araújo, F.R.F.N., Rocha, A.: Reactivation of Chagas' disease in patients with AIDS: Report of three new cases and review of the literature. Clin. Infect. Dis. **25**(6), 1397–1400 (1997)
9. Vaidian, A.K., Weiss, L.M., Tanowitz, H.B.: Chagas' disease and AIDS. Kinetoplastid Biol. Dis. **3**(1), 2 (2004)
10. World Health Organization: Sustaining the drive to overcome the global impact of neglected tropical diseases: Second WHO report on neglected diseases. Geneva (2013)
11. Oppenheim, A.N.: Questionnaire Design Interviewing and Attitude Measurement. Bloomsbury Publishing, London (1992)
12. Leffingwell, D., Widrig, D.: Managing Software Requirements: A Use Case Approach. Pearson Education, London (2003)
13. Sommerville, I., Kotonya, G.: Requirements Engineering: Processes and Techniques. John Wiley, New York (1998)
14. Osborn, A.F.: Development of Creative Education. Creative Education Foundation, Buffalo (1961)
15. Buzan, T.: Mind Mapping. Pearson, New York (2006)
16. Sommerville, I.: Software Engineering, 9th edn. Addison-Wesley, New York (2010)
17. Object Management Group: UML 2.0 superstructure specification (2004)
18. Warmer, J.B., Kleppe, A.G.: The Object Constraint Language: Precise Modeling with UML. Addison-Wesley, Boston (1998)
19. Winter, R., Strauch, B.: A method for demand-driven information requirements analysis in data warehousing projects. In: Proceedings of the 36th Annual Hawaii International Conference on System Sciences, p. 231. IEEE (2003)
20. Paim, F.R.S., de Castro, J.F.B.: DWARF: an approach for requirements definition and management of data warehouse systems. In: Proceedings of the 11th IEEE International Requirements Engineering Conference, pp. 75–84. IEEE (2003)
21. Connelly, R., McNeill, R., Mosimann, R.: The Multidimensional Manager: 24 Ways to Impact Your Bottom Line in 90 Days. Cognos, Canada (1999)
22. Valéry, P.: Notre destin et les lettres (1937)
23. Pedersen, T.B.: Multidimensional modeling. In: Encyclopedia of Database Systems. Springer, Heidelberg, pp. 1777–1784 (2009)
24. Lenz, H.J., Shoshani, A.: Summarizability in OLAP and statistical data bases. In: Proceedings of the 9th International Conference on Scientific and Statistical Database Management, pp. 132–143 (1997)
25. Maté, A., Trujillo, J.: A trace metamodel proposal based on the model driven architecture framework for the traceability of user requirements in data warehouses. Inf. Syst. **37**(8), 753–766 (2012)
26. Toranzo, M., Castro, J., Mello, E.: Uma proposta para melhorar o rastreamento de requisitos. In: Proceedings of the Workshop em Engenharia de Requisitos, pp. 194–209 (2002)
27. Britos, P., Dieste, O., Garcia Martinez, R.: Requirements elicitation in data mining for business intelligence projects. In: Avison, D., Kasper, G.M., Pernici, B., Ramos, L., Roode, D. (eds.) Advances in Information Systems Research, Education and Practice. IFIP, vol. 274, pp. 139–150. Springer, US (2008)
28. World Health Organization: Investing to overcome the global impact of neglected tropical diseases: Third WHO report on neglected tropical diseases. Geneva, Switzerland (2015)

Multi-perspective Analysis of Mobile Phone Call Data Records: A Visual Analytics Approach

Gennady Andrienko[1,2(✉)], Natalia Andrienko[1,2], and Georg Fuchs[1]

[1] Fraunhofer Institute IAIS, Schloss Birlinghoven,
53757 Sankt Augustin, Germany
{Gennady.Andrienko,Natalia.Andrienko,
Georg.Fuchs}@iais.fraunhofer.de
[2] City University London, Northampton Square, London EC1V 0HB, UK

Abstract. Analysis of human mobility is currently a hot research topic in data mining, geographic information science and visual analytics. While a wide variety of methods and tools are available, it is still hard to find recommendations for considering a data set systematically from multiple perspectives. To fill this gap, we demonstrate a workflow of a comprehensive analysis of a publicly available data set about mobile phone calls of a large population over a long time period. We pay special attention to the evaluation of data properties. We outline potential applications of the proposed methods.

Keywords: Visual analytics · Mobility data · Call data records

1 Introduction

Nowadays, huge amounts of movement data describing changes of spatial positions of discrete mobile objects are collected by means of contemporary tracking technologies such as GPS, RFID, and positions within mobile phone call records. Extensive research on trajectory analysis has been conducted in knowledge discovery in databases [1], spatial computing [2], and moving object databases [3]. Automatically collected movement data are semantically poor as they basically consist of object identifiers, coordinates in space, and time stamps. Despite that, valuable information about the objects and their movement behavior, as well as about the space and time in which they move can be gained even from such basic movement data by means of analysis [4].

Movement can be viewed from multiple perspectives: as consisting of continuous paths in space and time [5], also called trajectories, or as a composition of various spatial events [6]. Movement data can be aggregated in space, enabling identification of interesting places and studying their activity characteristics, and by time intervals, enabling similarity analysis of situations comprising different time intervals as well as detection of extraordinary events.

For the most comprehensive analysis of movement data, the analyst would look at the data from all perspectives: mover-oriented, event-oriented, space-oriented, and time-oriented. However, data properties often limit possible directions of analysis.

© Springer International Publishing Switzerland 2015
E. Zimányi and R.-D. Kutsche (Eds.): eBISS 2014, LNBIP 205, pp. 39–59, 2015.
DOI: 10.1007/978-3-319-17551-5_2

In this study, we consider the D4D fine resolution call data records (CDR) set of Ivory Coast [7] from multiple perspectives. To set the scope we first evaluate the properties of the data that restrict potentially applicable movement data analysis methods (Sect. 2). A first analysis step is to study spatio-temporal patterns of calling activities at multiple resolutions of time. To this end we apply spatio-temporal aggregations by antennas, counting number of calls per day (Sect. 3) and per hour (Sect. 4). To further identify different kinds of activity neighborhoods and to study their spatial distribution we then characterize antennas by feature vectors of hourly activities within a week and cluster them by similarity of the feature vectors (Sect. 5). In order to identify peak events – i.e., time intervals during which extraordinarily large number of people made calls in one location simultaneously - we compare time series comprising counts of distinct phone users per time interval and antenna (Sect. 6). This procedure allows us to identify large-scale events that, possibly, happened in the country. We use trajectories of mobile phone subscribers for reconstructing flows between major towns and between activity regions of the country (Sect. 7). Finally, we make an attempt at semantic interpretation of individuals' personal places, such as home and work locations, based on these user trajectories (Sect. 8). We conclude this paper with an outline of a general procedure of data analysis from multiple perspectives (Sect. 9) and a short discussion on the results and possible directions for further work.

2   Evaluating Data Properties

In analyzing movement data, it is important to take into account the following properties [14]:

- Temporal properties:
  - temporal resolution: the lengths of the time intervals between the position measurements;
  - temporal regularity: whether the length of the time intervals between the measurements is constant or variable;
  - temporal coverage: whether the measurements were made during the whole time span of the data or in a sample of time units, or there were intentional or unintentional breaks in the measurements;
  - time cycles coverage: whether all positions of relevant time cycles (daily, weekly, seasonal, etc.) are sufficiently represented in the data, or the data refer only to subsets of the positions (e.g., only to work days or only to daytime), or there is a bias towards some positions.
- Spatial properties:
  - spatial resolution: the minimal change of position of an object that can be reflected in the data;
  - spatial precision: whether the positions are defined as points (by exact coordinates) or as locations having spatial extents (e.g. areas). For example, the position of a mobile phone call is typically a cell in a mobile phone network;
  - spatial coverage: are positions recorded everywhere or, if not, how are the locations where positions are recorded distributed over the studied territory (in terms of the spatial extent, uniformity, and density)?

- Mover set properties:
  - number of movers: a single mover, a small number of movers, a large number of movers;
  - population coverage: whether there are data about all movers of interest for a given territory and time period or only for a sample of the movers;
  - representativeness: whether the sample of movers is representative, i.e., has the same distribution of properties as in the whole population, or biased towards individuals with particular properties.
- Data collection properties:
  - position exactness: How exactly could the positions be determined? Thus, a movement sensor may detect an object within its range but may not be able to determine the exact coordinates of the object. In this case, the position of the sensor will represent the position of the object in the data;
  - positioning accuracy, or how much error may be in the measurements;
  - missing positions: in some circumstances, object positions cannot be determined, which leads to gaps in the data;
  - meanings of the position absence: whether absence of positions corresponds to stops, or to conditions when measurements were impossible, or to device failure, or to private information that has been removed.

The provided data set [7] comprises a total of 55,319,911 CDRs distributed over ten individual chunks of between 4.8 and 6.5 million records, each corresponding to a set of two-week time intervals. Of these, 47,190,414 CDRs are associated with one of the 1,214 antennas and thus be referenced by the corresponding antenna's geographic coordinates. CDR temporal references are given with minute accuracy (i.e., seconds were suppressed) ranging from December 5, 2011 till April 22, 2012. Aggregation of geo-referenced calls by days (Fig. 1) shows that some days (e.g. March 24, 2012) have much less number of calls than neighboring days. This observation suggests that quite many call activities are missing in the database, especially in April 2012. In addition, 8,129,497 calls refer to unknown antennas (id = −1), with maximal count 166,621 calls on April 1, 2012. Because these CDRs could not be geo-located and thus not related to other calls originating from the same location they were ignored during data import.

The figure also suggests obvious call peak patterns at New Year, Easter, and, to some extent, at Christmas 2011. Other peaks correspond to public holidays like The Day after the Prophet's Birthday (Sunday, February 5, 2012) and Post African Cup of Nations Recovery (Monday, February 13, 2012)[1].

Wikipedia[2] suggests that religion in Ivory Coast remains very heterogeneous, with Islam (almost all Sunni Muslims) and Christianity (mostly Roman Catholic) being the major religions. Muslims dominate the north, while Christians dominate the south. Unfortunately, the amount of data available for the northern part of the country does not allow comparison of patterns in respect to religious holidays.

---

[1] Public holidays in Ivory Coast in 2012:
http://www.asaralo.com/index.php?option=com_content&view=article&id=2367:public-holidays-in-cote-divoire&catid=160:african-public-holiday&Itemid=2598.

[2] http://en.wikipedia.org/wiki/Ivory_Coast#Religion.

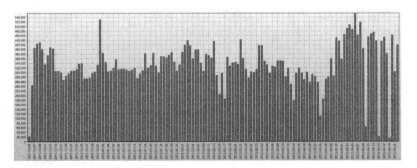

Fig. 1. Daily counts of calls.

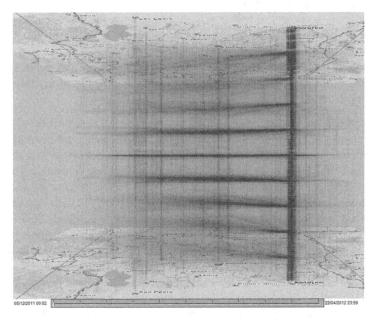

Fig. 2. Space-Time Cube displaying the full 20-week data set of CDRs integrated into trajectories (sequence of calls with the same user id) with time increasing from bottom to top of the cube. Besides expected daily cycles e.g. in the area of Abidjan one can spot missing days (near the top), and very clearly the distinct pattern of bi-weekly "false trip" movement caused by re-assigning user IDs to different mobile phone users in other parts of the country between data chunks.

A considerable constraint in terms of mobility pattern analysis and semantic interpretation (Sects. 7 and 8, respectively) arises from the anonymization procedure applied to the data [7]. Each of the 10 data chunks is a subset of 50,000 distinct mobile phone subscribers tracked over 2 weeks. User IDs associated with each CDR are obviously not real, traceable customer IDs but rather consecutive integer numbers. And while a given user ID is unique with respect to one data chunk, integers are reused (i.e., the counter was reset) between different chunks. This means that it is not possible to analyze movement patterns or flows over periods exceeding two weeks, or generally cover time intervals distributed over multiple chunks (compare Fig. 2).

Moreover, a check for repeated combinations of user ID and time stamp produced 5,225,989 pairs that occurred 12,861,168 times in the database. The duplicates have been removed. This operation thus reduced the number of geo-referenced CDRs in the database by about 25 %.

3 Assessing Daily Aggregates for Antennas

We have aggregated the remaining CDRs by antennas and days, producing daily time series of calls for each of the 1,214 antennas. Figure 3 presents an overview of their statistical properties. The upper part of the image shows the call counts' running average line (in bold) and dynamics by deciles (grey bands, min = 0, 10 %, 20 %, ..., 90 %, max = 5584) over time. Vertical lines correspond to weeks. The lower part of the image uses segmented bars to represent distribution of antennas categorized by their daily call counts. The darkest blue denotes absence of any calls at those antennas; blue colors correspond to intervals from 1 to 50 calls per day, yellow represents 50–100 calls, orange and reds – more than 100 calls.

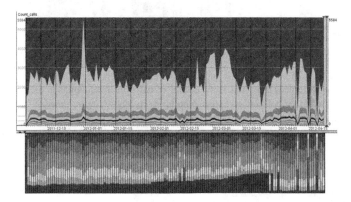

Fig. 3. Top: dynamics of deciles of counts of call per antenna distributions. Bottom: daily proportions of antennas with N calls in intervals of 0 (darkest blue), 1..10, 10..50, 50..100 (yellow), 100..200, 200..500, 500..1000, and more than 1000 (darkest red) per day. Note that in the upper image, corresponding interval boundaries are indicated in the scale to the left (Colour figure online).

We can make the following general observations:

- Too few data records on Dec 5, 2011 even though CDR time stamps for that day cover the entire 24 h period.
- Gradual increase of counts of antennas without activity (0 calls per day) from Dec 6, 2011 till March 27, 2012.
- Several days with missing data on many antennas (March 29, April 1, April 10, April 15, April 19).

44   G. Andrienko et al.

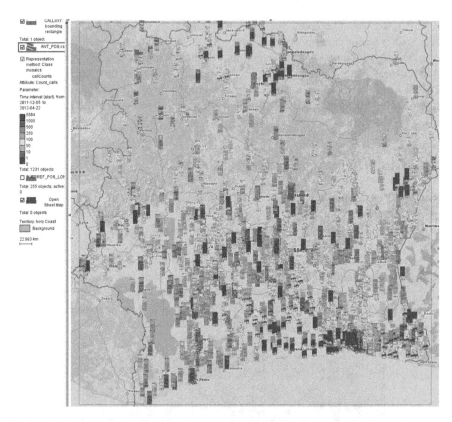

Fig. 4. Mosaic (segmented) diagrams show counts of calls for all antennas in the whole country. Counts are represented by colored segments ranging from blue (0 calls) through yellow (50..100 calls) to red (more than 1,000 calls). Diagram rows correspond to weeks (top to bottom – from week 1 to week 20) and columns to days of week (left to right: from Monday to Sunday) (Color figure online).

- Absence of typical weekly patterns with different amounts of calls at working days and weekends.

These general observations do not reflect the geographic distribution of patterns. To take the geography into account, we represent the call counts on maps by mosaic diagrams. A mosaic diagram consists of a pixel grid with each pixel representing one day's call count by color, using the same color coding as in Fig. 3. The pixels are arranged in 2D as in a calendar sheet: columns correspond to days of week (from Monday to Sunday, from left to right) and rows correspond to weeks (from 1 to 20, from top to bottom). Figure 4 shows the entire country, Fig. 5 a close-up of the region of the towns Abidjan and Abobo. The large consecutive sections of dark blue colors in many diagrams suggest that the data contain systematically missing portions. In particular, data are completely unavailable 12–14 weeks for many antennas in the northern part of the country, and for more than 16 weeks in the southern part of Abidjan.

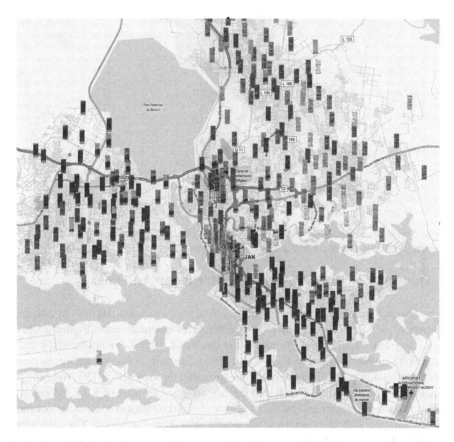

Fig. 5. Close-up view of the region of the towns Abidjan and Abobo. The mosaic diagrams are encoded in the same way as in Fig. 4 and use the same color coding.

Another observation is that all columns in the diagrams look quite similar. This is very different from mobile phone usage patterns observable in Europe and the USA where weekends differ strongly from working days in terms of calling counts. There, calls from the downtown areas of large cities are quite rare on Saturdays and Sundays in comparison to weekdays. We cannot find such patterns in the D4D data set. This suggests that the life style and temporal organization of economic activities in Ivory Coast differ significantly from those cultural regions. Therefore a straightforward application of analysis methods developed primarily for European countries is not valid.

One more complexity of the data is caused by the data sampling and anonymization procedures [7]. For each two-week period, a subset of 50,000 customers has been selected. It is not guaranteed that the subsets represent population samples with similar demographic and economical characteristics. Indeed, clustering days by feature vectors comprising counts of calls at each antenna, followed by assigning colors to clusters by similarity [8] clearly demonstrates the dissimilarity of patterns in consecutive two-weeks periods (Fig. 6). Additionally, this display also does not give any evidence of differences between week days and weekends.

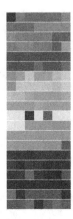

Fig. 6. Similarity of situations during 7 days ×20 weeks, represented by assigning colors to segments of the diagram according to the cluster the corresponding day belongs to.

4   Analyzing Hourly Aggregate Patterns for Antennas

Taking into account the properties of the data, we decided to aggregate calls by antennas for hours of day and days of week, irrespectively of calendar dates. Figure 7 shows mosaic diagram maps for two locations, the country's capital (Yamoussoukro) and a port town (San Pedro). Like in Figs. 4 and 5, the diagrams consist of segments representing call counts by colors, from dark blue (no calls) through yellow (50–100 calls per hour) to red. The segments of each diagram are arranged by days of week (Monday to Sunday from left to right) and by hours of day (from 0:00 on top to 23:00 at bottom).

One can see different temporal signatures of calling activities. Thus, in some antennas calls are more frequent at evening times, some have uniform distribution of call counts during daytime hours, while yet others have similar distributions at morning and evening times etc. However, the total amounts of calls differ significantly from one antenna to another, thus making direct comparison and grouping quite difficult.

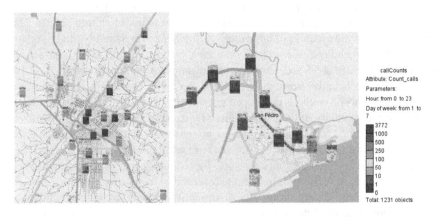

Fig. 7. Mosaic diagrams show hourly absolute counts of calls for 7 days of week (by columns, from Monday to Sunday) and 24 h of day (from 0:00 to 23:00) in Yamoussoukro and San Pedro.

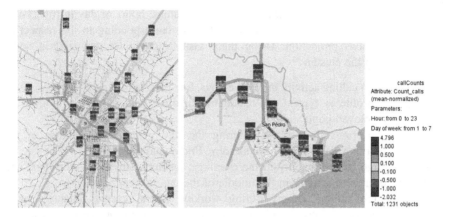

Fig. 8. Similarly to Fig. 7, mosaic diagrams show hourly show counts of calls for 7 days of week (by columns, from Monday to Sunday) and 24 h of day (from 0:00 to 23:00) normalized by average count per antenna in Yamoussoukro and San Pedro.

To compensate for different amounts of calls at different antennas, we have applied normalization to each time series by its own mean and standard deviation values, see Fig. 8. The resulting images convincingly demonstrate that there exist distinct patterns of hourly calling activities at different antennas. Moreover, these patterns tend to be clustered in geographical space. For example, almost all antennas in the outskirts of Yamoussoukro are characterized by dominant evening call pattern, while in the city centre calls are distributed uniformly during day. There are only few evidences of different calling activity patterns on Saturdays and Sundays (i.e., in the two rightmost columns of the diagrams) in comparison to working days. One such example can be found in the southern part of San Pedro, and some others in the southern part of Yamoussoukro.

5   Clustering Antennas by Similarity of Hourly Aggregate Patterns

Visual inspection and comparison of mosaic diagrams has limited applicability. We can perform it for selected cities and regions, but can't apply systematically for the whole country. Instead, we can apply clustering of antennas according to mean-normalized hourly activity profiles over week. We've used k-Means and varied the desired number of clusters from 5 to 15, the most interpretable results have been obtained with N = 7. Lower number of clusters mixes several behaviors, while large counts extract small clusters with too specific behaviors.

The results are presented in Fig. 9. Seven time graphs show profiles of the 7 clusters for 7 days of week. Centroids of the clusters have been projected onto the 2d plane by Sammons mapping [9] (middle left). Following the ideas of [10], colors have been assigned to the clusters according to these 2D positions, thus reflecting relative cluster similarities. The representative feature vector of the cluster centroids are presented by mosaic diagrams (middle right, days of week in columns, hours of day in rows, similarly to Figs. 7 and 8), with their placement again corresponding to the respective centroid's Sammons projection. Using these visualizations, we can suggest some interpretations to the clusters:

Cluster 1:   High calling activity in the evenings, irrespective of the day of week. Such a profile is typical for residential districts with a high proportion of employed population.

Cluster 2:   Uniform calling activity during the day, with some increase in the morning on Monday, Wednesday, Friday and Saturday.

Cluster 3:   High calling activity in the evenings, medium activity in mornings, and decreased activity in the middle of the day (except Sundays).

Cluster 4:   High calling activity during working hours (except Sundays), with extremes in mornings. Such a profile is typical for business districts.

Cluster 5:   Very low calling activity, with only small differences between day and night. This is quite typical for unpopulated areas and for antennas masked (in terms of call handling) by neighboring antennas.

Cluster 6:   Similar to cluster 3, however with a less prominent evening pattern but more prominent morning pattern, and increased activity on Saturdays and Sundays.

Cluster 7:   Similar to clusters 3 and 6, but with decreased activity on Sundays.

Our experience of analyzing mobile phone usage data in different countries suggests that cluster 1 corresponds to residential districts with high proportion of regularly employed population, in other words, people having fixed out-of-home work schedules, and that cluster 4 represents business districts. We guess that cluster 2 either represents regions with a mix of residential and business land use, or businesses with irregular schedules. Major transportation corridors (main roads, railways) can be characterized by similar temporal patterns, too. Clusters 3, 6 and 7 may represent mostly residential areas with partly employed population, or population with flexible work schedule.

The three maps at the bottom of Fig. 9 show, from left to right, the spatial distribution of the clusters for the whole country, its southern part, and the city of Abidjan, respectively. We can observe that our possible interpretations correspond to geographical patterns.

6 Peak Detection from Hourly Time Series at Antenna Level

Besides examining regular, everyday-life activity patterns we further want to detect interesting events that attracted many people. For this purpose, we need to count the number of different people per antenna cell and time unit (rather than the total number of calls/CDRs as used in the previous sections). It should be noted again that data have been provided in 2-weeks portions with repeated user IDs across the different portions, therefore limiting time intervals eligible for such analysis in this particular data set due to the inability to distinguish users between data chunks.

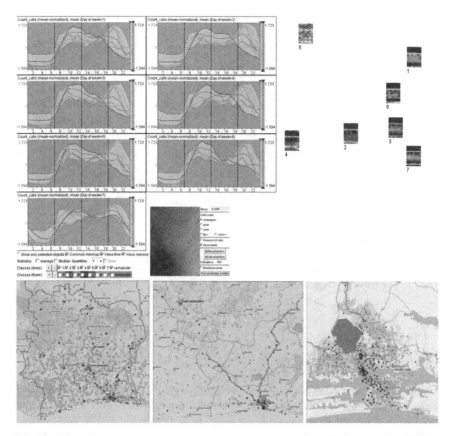

Fig. 9. Normalized temporal signatures of antennas are used for defining 7 clusters by k-Means. Time graphs in the upper panel show profiles of these clusters during 7 days of week. Colors are assigned to the clusters according to positions of cluster centroids in Sammons mapping (middle). Representative activity profiles for the clusters are shown by 2D mosaic diagrams in the yop-right. The maps at the bottom show spatial distributions of the clusters for the whole country (left), south-west part (center) and the region around Abidjan (right).

We focus our further analysis on trajectories (sequences of positions) of different users during last two weeks of the data set. This is the only period that contains rather complete geographic coverage, see Sect. 3 for details. For each distinct antenna we have computed hourly counts of distinct user IDs active at this antenna. These counts roughly represent the presence of people in antenna cells. If a person made several calls from the same antenna, we assume that he did not move away between the calls. It should be noted that this assumption may be incorrect in some cases, in that people may transition out of an antenna's cell and back without making a call at another antenna in the meantime.

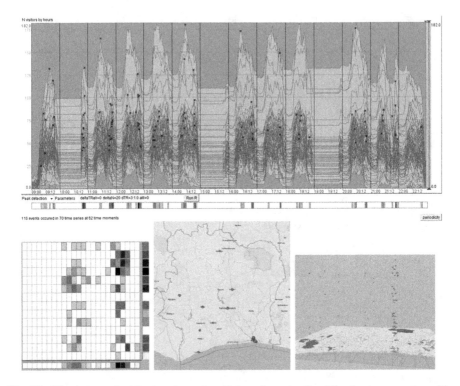

Fig. 10. The time graph at the top shows time series of counts of mobile phone users grouped by antennas, at 1 h resolution. Peaks with magnitude of at least 20 users over 3 h intervals are marked by yellow crosses. Counts of peaks are shown in 2d periodic event bar at the bottom-left. Positions of peak events are shown on the map of the country in the bottom-center map and in the space-time cube at bottom-right.

Figure 10 (top) shows a time graph with a selection of time lines corresponding to antennas. Straight horizontal lines on April 10, April 15 and April 19 correspond to missing data that we already identified earlier in Figs. 1 and 3. To find unusual concentrations of people at antennas, we have searched for peaks of averaged presence magnitudes exceeding 20 distinct peoples over a sliding, 3 h time window [11]. The appropriate parameters for magnitude threshold and time window width have been defined using a sensitivity analysis procedure as suggested by [12]. In particular, the time graph in Fig. 10 (top) only shows lines for those antennas that exhibit at least one such peak event. The horizontal event bar immediately below the time graph shows the counts of events over time. The 2D periodic event bar in Fig. 10 (bottom left) shows counts of peak events per 24 h of day (columns) and 14 days of two weeks (rows). The map (bottom-center) and space-time cube (bottom-right) show spatial and spatio-temporal distributions of peak events.

We can observe that peak events are frequent in the middle of the day and early in the evening. There are only few exceptions. Thus, several peak events happened during the 15:00–16:00 h interval on Monday and Fridays of the 1st week, and late in the evening of Saturday of the 2nd week. By clicking the corresponding segment of the periodic event bar, we select the corresponding antennas and time series (see Fig. 11). We can see that these peaks happened in 4 different towns in different parts of the country. The time series profiles for those regions indicate that these peaks are rather unusual. We guess that some kind of connected public events happened simultaneously in these regions.

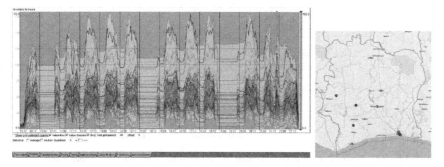

Fig. 11. Peaks that happened at 21:00 on the 2nd week's Saturday and their containing time series are highlighted in the time graph (left). Simultaneously, their positions are marked on the map (right).

It is interesting to relate the magnitude of peaks with the maximal values of the time series. We found two extreme cases of time series with peaks of more than 20 people contained in time series with maximum (peak) values of about 40 but average daily values of only about 10..15 people (Fig. 12). Both events happened in Abidjan. Probably, some local events happened at about 10:00 on Monday and at 21:00 of Thursday in these locations.

We found that peak events happened in almost all major towns of the country. To get a flavor of mobility of mobile phone users in Ivory Coast, we outlined areas around the peak events and then calculated counts of direct transitions between these locations,

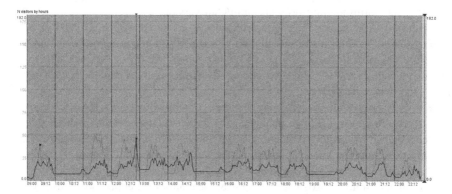

Fig. 12. Peaks on Monday morning (yellow cross) and Thursday evening (green cross) are shown on top of two time series with otherwise usually low presence of calling activities. Both peaks have happened in Abidjan (Colour figure online).

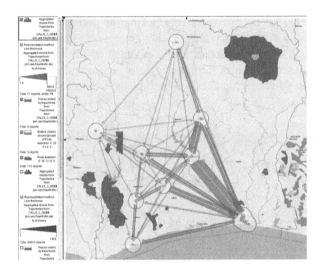

Fig. 13. Flows between regions that correspond to peaks in people presence.

see Fig. 13. The thickness of the arrows reflects the magnitude of flows between the corresponding places during the two-week period. This map shows us the strength of connections between locations of different activities.

7   Analysis of Flows

To explore the mobility flows more systematically, we have applied a method for generalization and summarization of trajectories [13] to the phone user trajectories over last two weeks. The method extracts so-called characteristic points of trajectories, aggregates these points in space with a desired resolution, and finally uses the medoids of the resulting spatial clusters as seeds for generating a Voronoi tessellation of the territory. The method simplifies trajectories while minimizing their distortion with respect to the corresponding original, full-detail version.

Figure 14 (left) shows the original trajectories rendered with high level of transparency (about 99 %). This representation gives us a hint about major flows, but does not allow quantifying them. Figure 14 (right) shows the flows between aggregated regions as well as the accumulated counts of distinct users recorded in each region during the two-week period.

We can observe the consistency between the flow maps in Figs. 13 and 14, respectively. However, the latter map uncovers more structural details. In particular, we can see a branch connecting Abidjan with the mid-eastern region of the country. There are only relatively few direct connections between Abidjan and Yamoussoukro, and fewer still between these two and towns in the northern part of the country. This indicates that despite the existences of several local airports, people mostly use ground transportation and make phone calls/send SMS during their lengthy trips. By contrast, air travel typically manifests itself as long-distance flows since the mobile phone is switched of or out of range during flight with no calls at intermediate antennas.

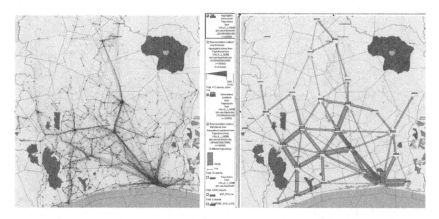

Fig. 14. All trajectories during last two weeks drawn as accumulation of semi-transparent lines (left). Trajectories are summarized by 28 aggregated regions (Voronoi polygons) of approximately 100 km radius. Flows between regions are represented by red arrows with flow magnitudes encoded in the arrow width. Counts of mobile phone owners registered in each area are shown by yellow bars (Colour figure online).

Further analysis (omitted here for space/time constraints) could allow us to identify temporal patterns of flows and assess usual travel times between different locations. We could also find frequent sequences of visited regions and assess the dynamics of such trips.

8   Semantic Analysis of Personal Places

To identify routine trips of people and to obtain interpretations of their personal places, we have applied the procedure proposed in [14] to a small subset of trajectories that are characterized by large numbers of calls in different locations. We have used a sample of the data consisting of 86 trajectories recorded during the last two weeks of the data period and with bounding rectangle diagonals exceeding 5 km. The total number of call records in this sample is 133,029. First, we have identified stops as sequences of consecutive calls that occurred within 30 min and a rectangular region of less than 500 m diagonal. Using these parameters extracted 7,149 stops. The stops have then been clustered by means of the density-based clustering method Optics [15], separately for each trajectory. Parameters have been chosen to group points having at least 5 neighbors within 500 m distance. Noise points not grouped into any cluster (1,300 points in total, or about 19 % of the set) have been excluded from subsequent analysis as they are assumed to represent infrequently visited locations. For each cluster the counts of calls have been aggregated for every hour of the day. This resulted in time series comprised of 24 one-hour intervals assigned to each cluster.

Figure 15 shows routine activity locations for a single person, id #548709, in space and time. The "blue" place to the south was attended only during day time. Most probably, this is the work place of this person. We guess that she has regular work with fixed working times. The "purple" place in the middle is visited less frequently and only

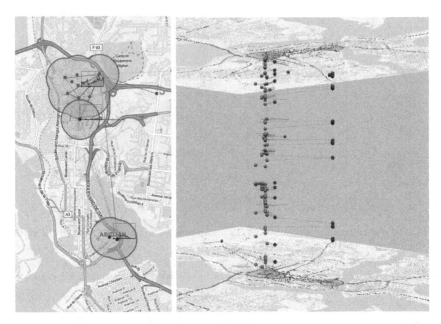

Fig. 15. Individual locations of repeated activities are shown by 500 m buffer polygons for subscriber #548709. Hourly temporal signatures (according to hours of day) are shown by time flow diagrams. Spatio-temporal positions of calls are shown in the space-time cube. Red dots represent home-based calls, blue dots correspond to the person's work place, and prurple to the primary location of her evening activities. Gray dots in the space-time cube represent irregular activities (Colour figure online).

during evening times. We guess this is a place of repeated social activities of the person. Finally, the location in the north is characterized by activities at any times, including night times (but less during the day). We interpret this place as the person's home.

Interpretation of semantically meaningful personal places can be automated. For example, we can compute similarities of all 24-hour feature vectors to a selected vector corresponding to "work" activities, see Fig. 16. This map shows locations and temporal signatures of places that can be interpreted as work locations for different mobile phone owners. Partial dissimilarity of the temporal profiles suggests different working hours. For example, some persons seem to be less active at lunch times. Spatial clustering of several people's "work" places suggests concentrations of business locations in the city.

A better quality of semantic interpretation could be achieved if CDRs included times and positions of both the beginnings and ends of calls. In this case it becomes possible to distinguish stationary calls from calls on move, and to estimate movement speeds during the latter.

By applying the described procedure systematically to all subsets of the data and matching routine activity locations of persons in different subsets, it might further be possible to link partial trajectory corresponding to the same person in different data chunks (see Sect. 2). However, such re-integration may be harmful in terms of personal privacy [16].

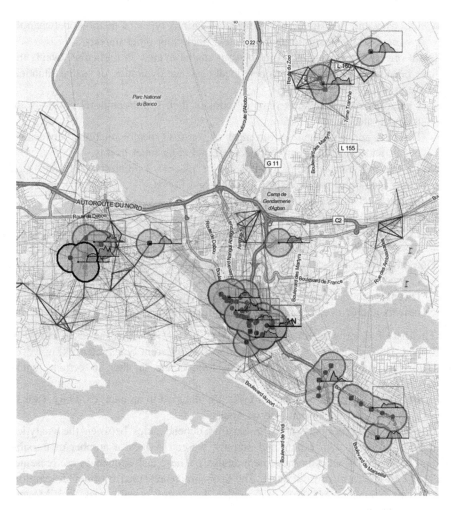

Fig. 16. Locations from other trajectories characterized by temporal profiles similar to that was identified as work in Fig. 15.

9   A General Procedure of Analysis

In this section, we attempt to outline a general procedure for analyzing movement data from multiple perspectives. For the most comprehensive analysis of movement data, the analyst would look at the data from all perspectives: mover-oriented, event-oriented, space-oriented, and time-oriented. Such an analysis would include the following groups of tasks:

- Mover-oriented tasks dealing with trajectories of movers:
    - Characterize trajectories as units in terms of their positions in space and time, shapes, and other overall characteristics.
    - Analyze the variation of the positional attributes in space and time.

- – Discover and investigate occurrences of various types of relations between the movers and the spatio-temporal context, including other movers.
- Event-oriented tasks dealing with relevant spatial events, in particular, events that have been extracted from trajectories, local presence dynamics, or spatial situations in the process of the analysis:
  - – Characterize the relevant events in terms of their spatio-temporal positions and thematic attributes.
  - – Discover and investigate occurrences of various types of relations between the events and the spatio-temporal context, including other events.
- Space-oriented tasks dealing with a set of places of interest (POI) and local dynamics (temporal variations) of presence and flows:
  - – Define a set of relevant POI.
  - – Characterize the POI in terms of the local presence dynamics.
  - – Characterize binary links between the POI in terms of the flow dynamics.
  - – Discover and investigate temporal and ordering relations between the POI.
- Time-oriented tasks dealing with a set of time units and respective spatial situations:
  - – Characterize the time units in terms of the spatial situations.
  - – Discover and characterize the relations between the time units imposed by movers and/or events, in particular, similarity and change relations.

This list of tasks is not meant to specify any order in which the tasks should be performed. During the process of analysis, tasks of different types intermix; however, they do not intermix fully arbitrarily but follow one another in certain logical sequences.

It is not necessary that all types of tasks are included in an analysis. Only a subset of tasks may be relevant to the analysis goals.

Based on our experience and the existing dependencies between the analytical methods in terms of their inputs and outputs, we can suggest a number of possible rational sequences of tasks in movement analysis. These task sequences are presented in Fig. 17 in the form of flow chart. The tasks are represented by brief descriptions preceded by characters M, E, S, or T, which denote the possible task foci: Movers, Events, Space, and Time.

Although the graph specifying the possible task sequences has a single root node, it does not mean that any analysis must begin with the task "Analyze trajectories as units" represented by this node. For a particular application, the characteristics of trajectories as units may be of no interest but analysts may be interested first of all in the positional attributes or in relations of movers to the context or in aggregated movement characteristics over a given territory. Furthermore, the analysis may initially focus on spatial events, in particular, when the movement data are originally available in the form of spatial events rather than trajectories, as, for example, data from Flickr or Twitter or data about mobile phone use. In the flow chart, the nodes where the analysis can start are marked by grey background.

It is also not necessary that the analysis ends only when one of the terminal nodes is reached and the respective task fulfilled. The analysis may end in any intermediate node when the application-relevant analysis goals are achieved. The analysis may also continue by switching to another branch. In particular, there are two terminal nodes

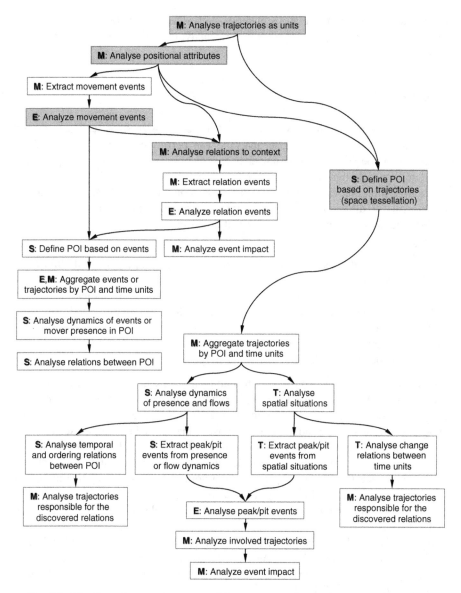

Fig. 17. The flow chart represents possible sequences of tasks in movement analysis.

labelled "M: Analyze trajectories responsible for the discovered relations" (where relations between POI or time units are meant). Here it is assumed that a subset of trajectories is selected for which the analysis is done starting from the root node of the flowchart and following the left branch.

Hence, there is no unique analysis procedure that needs to be followed in all cases but there are many possible procedures, where the steps are chosen depending on the application-specific analysis goals and ordered according to the dependencies between the inputs and outputs of the analysis methods. Nevertheless, the possible paths through the flow chart in Fig. 17 specify a set of generic analytical procedures that can be useful in multiple applications.

10 Conclusions

In this paper, we report on analysis results of a medium-size set of call data records referring to antenna positions. The analysis was performed with V-Analytics – a research prototype integrating visual analytics techniques for spatial, temporal and spatio-temporal data that our group develops since the mid-90 s [17]. We considered the data from multiple perspectives, including views on locations of varying resolution, time intervals of different length and hierarchical organization, and trajectories. We detected a number of interesting patterns that could facilitate a variety of applications, including

- Reconstructing demographic information (to replace expensive and difficult to organize census studies).
- Reconstructing patterns of mobility (to enhance transportation studies).
- Identifying places of important activities (for improving land use and infrastructure).
- Identifying events (for improving safety and security).
- Detecting social networks (for marketing purposes).

While in some cases we considered the complete data set, we had to restrict parts of our analysis to the last two weeks of the provided data due to undesirable properties (namely, missing, incomplete or duplicate data records for several key regions for a large portion of the time period). However, most of the applied techniques scale (or can be scaled up conceptually) for much larger data sets. Some kinds of analysis that we planned to perform were simply impossible due to the data fragmentation into chunks with duplicate user IDs. For example, we were not able to build predictive models of people's presence and mobility [18], as data for longer time periods are needed. We also did not search for interaction patterns between people and did not try to detect social networks.

Limitation caused by data quality could be relaxed by joining the provided data set with data from publicly available sources such as Flickr and Twitter in future work. Textual aggregates of activity records could greatly facilitate understanding and deeper semantic interpretation of the data.

References

1. Giannotti, F., Pedreschi, D. (eds.): Mobility, data mining and privacy - geographic knowledge discovery. Springer, Berlin (2008)
2. Laube, P.: Progress in movement pattern analysis. In: Gottfried, B., Aghajan, H.K. (eds.) BMI book of ambient intelligence and smart environments, vol. 3, pp. 43–71. IOS Press (2009)

3. Güting, R.H., Schneider, M.: Moving objects databases. Morgan Kaufmann, Burlington (2005)

4. Andrienko, G., Andrienko, N., Bak, P., Keim, D., Kisilevich, S., Wrobel, S.: A conceptual framework and taxonomy of techniques for analyzing movement. J. Vis. Lang. Comput. **22**(3), 213–232 (2011)

5. Hägerstrand, T.: What about people in regional science? Papers, vol. 24, pp.7–21. Regional Science Association (1970)

6. Andrienko, G., Andrienko, N., Heurich, M.: An event-based conceptual model for context-aware movement analysis. Int. J. Geogr. Inf. Sci. **25**(9), 1347–1370 (2011)

7. Blondel, V., Esch, M., Chan, C., Clerot, F., Deville, P., Huens, E., Morlot, F., Smoreda, Z., Ziemlicki, C.: Data for development: the D4D orange challenge on mobile phone data. http://arxiv.org/abs/1210.0137

8. Andrienko, G., Andrienko, N., Bak, P., Bremm, S., Keim, D., von Landesberger, T., Pölitz, C., Schreck, T.: A framework for using self-organizing maps to analyze spatio-temporal patterns, exemplified by analysis of mobile phone usage. J. Locat. Based Serv. **4**(3/4), 200–221 (2010)

9. Sammon, J.W.: A nonlinear mapping for data structure analysis. IEEE Trans. Comput. **18**, 401–409 (1969)

10. Andrienko, G., Andrienko, N., Bremm, S., Schreck, T., von Landesberger, T., Bak, P., Keim, D.: Space-in-time and time-in-space self-organizing maps for exploring spatiotemporal patterns. Comput. Graph. Forum **29**(3), 913–922 (2010)

11. Andrienko, G., Andrienko, N., Mladenov, M., Mock, M., Pölitz, C.: Discovering bits of place histories from people's activity traces. In: Proceedings IEEE Visual Analytics Science and Technology, pp.59–66. IEEE Computer Society Press (2010)

12. Andrienko, G., Andrienko, N., Mladenov, M., Mock, M., Pölitz, C.: Identifying place histories from activity traces with an eye to parameter impact. IEEE Trans. Vis. Comput. Graph. **18**(5), 675–688 (2012)

13. Andrienko, N., Andrienko, G.: Spatial generalization and aggregation of massive movement data. IEEE Trans. Vis. Comput. Graph. **17**(2), 205–219 (2011)

14. Andrienko, G., Andrienko, N., Bak, P., Keim, D., Wrobel, S.: Visual analytics of movement. Springer, Berlin (2013)

15. Ester, M., Kriegel, H.-P., Sander, J., Xu, X.: A density-based algorithm for discovering clusters in large spatial databases with noise. In: Second International Conference on Knowledge Discovery and Data Mining, pp.226–231. Portland, Oregon (1996)

16. Andrienko, G., Andrienko, N.: Privacy issues in geospatial visual analytics. In: Proceedings 8th Symposium on Location-Based Services, pp.239–246. Springer, Berlin (2011)

17. Andrienko, N., Andrienko, G.: Visual analytics of movement: an overview of methods, tools, and procedures. Inf. Vis. **12**(1), 3–24 (2013)

18. Andrienko, N., Andrienko, G.: A visual analytics framework for spatio-temporal analysis and modelling. Data Min. Knowl. Disc. **27**(1), 55–83 (2013)

From the Web of Documents to the Linked Data

Gabriel Képéklian[1][(✉)], Olivier Curé[2,3], and Laurent Bihanic[1]

[1] Atos France, Beauchamp, France
{gabriel.Kepeklian,laurent.Bihanic}@atos.net
[2] Sorbonne Universités, UPMC Univ Paris 06, UMR 7606, LIP6,
75005 Paris, France
olivier.cure@lip6.fr
[3] CNRS, UMR 7606, LIP6, 75005 Paris, France

Abstract. We start off this paper with a description of the evolution that led from the first version of the Web to the Web of data, sometimes referred to as the Semantic Web. Based on the "data > information > knowledge" hierarchy, we adopt a usage-based perspective. We then make explicit the structures of knowledge representation and the building blocks of the Web of data. In the next chapters, we show how RDF (*Resource Description Framework*) data are managed and queried. This allows us to describe an innovative data processing platform which produces linked data. We continue on ontologies, from their creation to their alignment that allows us developing the delicate question of reasoning. We can then conclude on research and development prospects offered by the linked data environment.

1 Some Historical Milestones to the Web of Data

As the main designer and visionary of the Web, Tim Berners-Lee had anticipated on the evolution from a Web of documents to a Web of data. In order to apprehend correctly the development and full dimension of this emerging Web and its promising research perspectives, it is important to clarify the notions of data, information and knowledge.

1.1 Data's Stance in the Early Web

As stated in [3], "Data are defined as symbols that represent properties of objects, events and their environment". Taken alone, data are not very useful. They have value when they are correctly interpreted.

In the first version of the Web, end-users were the only agents with the ability to provide such an interpretation. In fact, programs supporting the Web were manipulating data stored in complex data structures but were only delivering documents containing them. This motivated the Web of Documents designation. With the amount of Web pages rapidly increasing, it was necessary to organize things in order to help the search process. Along directories like *dmoz.org*, appeared search engines, e.g., *Altavista*, *Yahoo!*, *Google*. But the problem was still

© Springer International Publishing Switzerland 2015
E. Zimányi and R.-D. Kutsche (Eds.): eBISS 2014, LNBIP 205, pp. 60–87, 2015.
DOI: 10.1007/978-3-319-17551-5_3

the same: end-users were still being proposed documents out of search expressed in terms of few words. The search engines were not capable of interpreting the data contained in HTML documents.

We well know the confusion and errors that this entails. For instance, if one enters the word "Venus" in a search engine, what one can expect to get as a result? The planet, the Botticelli painting (i.e., The birth of Venus) or the tennis player (i.e. Venus Williams)? When the search possibilities remain lexical, it is not possible to take the context into account, and so to precise the field of knowledge, to know the freshness of the informations in the page, impossible to establish their reliability, to evaluate the sources, to compare with other available sources, etc.

Alone, data has no meaning. Data has a value only when placed in a precise context and when that context in itself gives sense or contributes to give one, that is to say, it provides access to meaning by building relations.

1.2   Information Emergence

The notion of information helps in decision-making or in problem solving. It is of a different order than the data. Information is a message that contains meaning, and whose inclusion will allow decision or action. Although well understood, this definition does not tell us how to achieve that goal.

It is important to understand that it is when we express something about a data that it becomes an information that we can share with people. The language allows locating the data in a statement; it places the data either as the subject of the sentence, or as the object. In "John eats an apple", we have two data: "John" and "apple". With the presence of the verb "eat", the sentence becomes information and the assembly makes sense. To move from data to information, we required a vocabulary.

At its origins, the Web was presented as a "global information space of linked documents". The fact that the links are between the documents is a main limitation especially due to the indescribable relationship between information and the document. How should we arrange the information to make a document? How to decompose a document? What is the nature of links between documents? As we know, these links between documents are purely technical, they have no semantic value. If two documents are linked, it is not possible to qualify the relationship between them.

It is well understood that there is information in the document. But the links are also rich in information. And in some cases, it can even be considered that it is these links that bring the most value and meaning.

1.3   The Advent of Knowledge

The information depends on two factors: the first deals with the data, these are facts emanating from sources, the second is communication between entities. For stakeholders to understand the communication, they must share the knowledge the information refers to.

Moreover, it is a fact often heard that we produce data in excess. And the fear that too much data kill information is not far. Certainly, the data deluge is real, the phenomenon of big data is not a fiction and mechanically causes an inflation of the same order of magnitude for the information. But if big data can lead to big information, big information doesn't lead to big knowledge.

Knowledge is a set of conceptual structures. This is the reason why the volume of information is uncorrelated with the knowledge that is expressed. For example, once we know a language, we are able to share all information with our neighbors.

Knowledge meets the requirement of a group to share information, understand it to take decisions or undertake actions, as we mentioned earlier in this section.

1.4   Summary

A. Liew noted that the usual definitions of data, information and knowledge had all one major flaw: "they are defined with each other, *i.e.* data in terms of information, information is defined in terms of data and/or knowledge, and knowledge is defined in terms of information. If we are just describing the interrelationships, that is all very well. However, with regard to definitions, this is a logical fallacy i.e. circular definitions or argumentations" [15].

It is precisely because there are relationships between these concepts, that a path leads from the data to the linked data. "Data are Elements of analysis (. . .) Information is data with context (. . .) Knowledge is information with Meaning" [4]. The knowledge is then used to understand, explain, decide and act.

2   The Structures of Knowledge Representation

At the heart of information architecture, classification techniques are particularly essential while growth of online data volume is accelerating. The ratio of what we are looking for to what we are not looking for is increasingly unfavorable in constant technology. The Semantic Web offers new approaches to tackle this problem. We will address them in a graduated manner.

2.1   Controlled Vocabulary

As the name suggests it, the purpose of controlled vocabularies is to control the set of terms that forms a vocabulary. The name of the days of the week is an example for understanding that this set is of finite dimension and can be exhaustively described. The list of species names is another example (see Fig. 1); but that will never be exhaustively defined because we keep discovering new species on a regular basis. These two examples show the character of referential lists of controlled vocabularies where the only relationship between the elements is that of belonging to the whole.

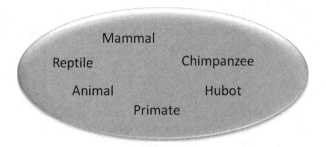

Fig. 1. Controlled vocabulary

In practice, controlled vocabularies often consider the following good practice: all words must be expressed in the same language. A set of names of the months of the year that would mix different languages, e.g., French and English, would serve no purpose. It should also be noted that the translation of a controlled vocabulary in another language can induce semantic variations that the structure cannot state. For example, the set of all the names given to the green color does not have the same cardinality in different languages. So, a good practice is to better characterize the controlled vocabulary.

The advantages of controlled vocabularies are the limitation of indexing disparity, of synonymy problems and of multiple or incorrect spellings.

2.2 Taxonomy

The need to build sub-groups in a controlled vocabulary, to organize terms into categories, leads to a tree-like structure which is quite natural, see Fig. 2. It becomes possible to place elements not only in respect to the whole, but also in relation to each other using hierarchical relations. These ordering relationships take into account classified realities like movie or musical genres, menus of an application, etc.

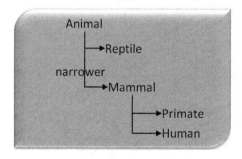

Fig. 2. Taxonomy

A taxonomy has several advantages. It is a classification system very simple to build, to understand and to exchange about. It provides a mapping of knowledge organized in levels of increasing details.

2.3   Thesaurus

The thesaurus is a structure which increases again the level of represented knowledge. It is often presented as a linguistic tool to overcome the ambiguities of natural language when defining shared concepts.

But unlike taxonomies, thesauri structure is rarely strictly hierarchical - otherwise we would be satisfied with taxonomies. Indeed, beyond the hierarchical relationships, a thesaurus allows using other properties to describe relationships between terms.

The most common properties are:

- BT = Broader Term (a symbol used in a thesaurus to identify the following terms as broader terms to the heading term);
- NT = Narrower Term (indicates one or more specific, or child descriptors)
- RT = Related Term ("See also")
- SN = Scope Note (if present, provides an explanation on using the given descriptor)
- TT = "Top Term" (identifies the "ancestor" term, the highest in the hierarchy, the highest using the relationship "BT")
- UF = Used For
- USE = "See" (refers the reader from a variant term to the vocabulary term).

Figure 3 provides a small example of a thesaurus using several of these properties. A thesaurus also allows stating a preference between several terms describing the same concept. Taxonomies do not permit such descriptions.

SKOS (*Simple Knowledge Organization System*) is an RDF vocabulary for modeling and defining thesaurus (and of course taxonomies). The UNESCO

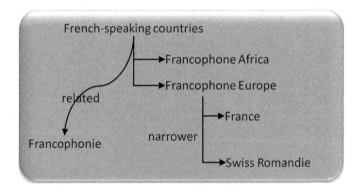

Fig. 3. Thesaurus

thesaurus[1] is a fine example that can be queried easily through a Web interface. It is very rich and continuously updated.

2.4   Ontology

Ontologies represent the higher degree of sophistication of knowledge structuring. Unlike previous approaches offering a limited number of relationships, it is possible to define new relationships in an ontology and thus to act on the language description of a domain of knowledge. An ontology is indeed a formal language, that is to say, a grammar that defines how the terms can be used together.

The aim consists in giving a unique meaning to concepts in order to avoid polysemy and to achieve consensus within a community of practice. Contrary to the approach of expert systems, an ontology describes a field of knowledge regardless of the uses that will be made.

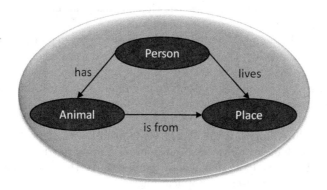

Fig. 4. Ontology

The ontology terms may be used to select the knowledge in which they are used. They can therefore be used as selection criteria, *i.e.*, viewpoints. The terms can be arranged, different perspectives on knowledge can be structured. In Fig. 4, concepts (Animal, Person, Place) are linked by relations (has, lives, is from). Note that the links are generally oriented and some cyclic graphs can be obtained. Many practical ontology languages are being represented using a logical formalism and will hence support some form of automatic reasoning (see Sect. 6 of this paper).

2.5   Summary

As we have seen, knowledge is mainly designed for sharing. And to share knowledge, they should be organized rationally. Besides the four structures presented here, there are other topics such as folksonomies - a set of tags organized as a classification system. The key is to find the right level of knowledge representation for the problem at hand. For example, if a thesaurus is sufficient, there

---

[1] http://databases.unesco.org/thesaurus/.

is no need to create an ontology since its construction is more complex and its manipulation is more involved.

3 Building Blocks for the Web of Data

We can consider that there are three building blocks to the Web of data: HTTP (*HyperText Transfer Protocol*), URI (*Uniform Resource Identifier*) and RDF. In this section, we present each of them and motivate the RDF data model against the popular relational one.

3.1 URI and HTTP

In RDF, URIs (actually IRIs: *Internationalized Resource Identifiers*, an extension of URIs that fully supports Unicode) identify resources. Hence, one of the key aspects when planning publishing information as linked data is to carefully define the URI naming policy.

Technically, URIs can be divided into two distinct sets depending on whether they can be dereferenced. Dereferenceable URIs use a scheme associated to a network protocol (`http`, `ftp`...). URIs using schemes not associated to a protocol (`urn`, `ark`...) can identify resources but require an additional component (a *resolver*) to support network access to the resources.

When publishing linked data, the protocol of choice is HTTP. This is one of the core building blocks of the World Wide Web (W3). It provides a simple and universal mechanism for accessing resources and retrieving representations (descriptions of a resource). That choice alone ensures that the data be widely accessible, from web browsers to applications.

One of HTTP core features is content negotiation. Upon access to the same URL, clients can request and obtain a view of the resource that best matches their needs or capacity to support a given output. For instance, one could retrieve a simple HTML page to be displayed in a browser or binary formats suitable for direct consumption by applications (picture, map, spreadsheet file, video...).

With RDF, it is also important to distinguish a resource from its representations. A resource is an abstract set of information that may (or may not) correspond to a real world object. Upon access (typically an HTTP request), the server will return a view of the resource, a representation. With HTTP, this representation is subject to content negotiation, *i.e.*, the client lists the types of representations (denoted as MIME (*Multi-Purpose Internet Mail Extensions*) types) it is ready to accept with, for each one a level of preference, and the server selects the best matching candidate among the representation the resource can offer.

The URI policy then dictates whether the server will directly return the selected representation in the response or instead redirect the client to a specific URL (Uniform Resource Locator). In this case, each representation of a resource has its own identity (URL). This allows for example, users to share or bookmark URLs that include both the reference of the resource and the type of representation to provide.

Several governmental organizations have started publishing guidelines to define and manage URIs in the context of Linked Data, for example INSEE in France [8] or Data.gov.uk in the UK [19]. These documents are good starting points for defining local URI policies.

3.2 Limits of the Relational Model

Before presenting the RDF data model, one can ask why we are not using an existing data model. As of today, most data are stored following the relational model [2]. Yet this model suffers from severe limitations that hinder consumption, sharing or reuse of these data. Although the relational model is not the only model for managing data, it is still the dominant one and most other existing models (hierarchical, key-value...) suffer from the same limitations.

The first limit comes from the separation between the structure of the data and the data themselves. A data corresponds to a column associated to a record in a table and can not be understood outside the context of the record. Moreover, the structure of the data is captured by the definition of the table, hence outside the data itself.

The structure of the data is also quite rigid and opaque:

- If a data is missing, a fictional NULL value shall be inserted;
- If multiple values occur for the same data, it is necessary to define a second column or use a separate table;
- There is no support for internationalized labels, these labels must be handled specifically, typically as a multi-valued field;
- Relations between records is not explicit but relies on usage of shared identifiers (foreign keys). The nature or intent of the relationship is never expressed, either in the date itself or in the structure (tables);
- The data structure is local. There is no standard way to convey this structure (and the relationships between data) along with the data themselves (exports).

The relational model does not provide any form of universal naming. Record identifiers are not normalized and are most of the time local to each table or, in the best cases, the local database.

As a consequence, database access and data extracts are specific to each product. There is no standard protocol for querying remote databases nor standard data formats to transfer data and their structure between databases from different providers.

So, there is a need for alternative data models that address these shortcomings. The RDF data model is the W3C response to these challenges in the context of the World Wide Web.

3.3 RDF

The objective of the W3 Consortium when it started designing RDF was to provide interoperability between applications that exchange machine-understandable information on the Web. Originally RDF was aiming at expressing metadata, i.e. "data about data", specifically to describe web resources.

RDF is not a data format. It is a syntax-independent model for representing named properties of resources and their values. A resource is identified by a URI. Values can hold either a literal (string, integer, date...) or a URI, in which case the property models a relationship between two resources. In the RDF data model, resources and values are nodes in a directed labeled graph, the properties being the arcs connecting nodes.

In RDF, each information is expressed as a statement of three elements, a triple: the resource (subject), the property (predicate) and the value (object). Each statement is independent and self descriptive. Hence, it is possible to extract as little information as needed from an RDF document without loosing context information.

In order to be able to transfer RDF statements between applications, serialization formats were required. The W3C has proposed a standard syntax, RDF/XML[2], but other formats exist and are commonly used: Turtle[3], Notation 3 (N3)[4], N-Triples[5] and more recently JSON-LD[6]. As the usage of RDF stores (triple stores) increased, new syntaxes appeared to support exporting whole named graphs, for example: Trig[7] and N-Quads[8].

4 Managing and Querying RDF Data

The ability to manage large RDF data sets is a fundamental prerequisite to the fulfillment of the Semantic Web vision. The W3C Data Activity[9] regroups a set of standards enabling the design of Knowledge Base Management Systems (KBMS). In this section, we present the storage and indexing approaches adapted to the RDF data model. Then, we present the SPARQL (*SPARQL Protocol and RDF Query Language*) query language which is mainly used to express queries over these systems. Finally, we present two original important notions that are specific to RDF and SPARQL, namely SPARQL endpoints and federated queries.

4.1 Triple Stores

RDF is a logical data model and hence does not impose any constraint on the physical storage approach. This opens a wide range of possibilities to design RDF database management systems, henceforth RDF Stores. There exists more than fifty different RDF stores which can be categorized as native or non-native [9].

---

[2] http://www.w3.org/TR/rdf-syntax-grammar/.
[3] http://www.w3.org/TR/turtle/.
[4] http://www.w3.org/TeamSubmission/n3/.
[5] http://www.w3.org/2001/sw/RDFCore/ntriples/.
[6] http://www.w3.org/TR/json-ld/.
[7] http://www.w3.org/TR/trig/.
[8] http://www.w3.org/TR/n-quads/.
[9] http://www.w3.org/2013/data/.

The non-native stores are based on an existing data management system. Typically, this can be a Relational DataBase Management System (RDBMS) and in that case, the RDF Stores benefits from many functionalities that have been designed over more than thirty years of research and development. Of course the original RDF graph-based data model has to be adapted to the relational model. The most popular approach consists in storing all RDF triples in a single relation containing only three columns, respectively one for the subject, property and object. Figure 5 presents an extract of such a triple table that stores items of a library. This simple approach presents several drawbacks. In particular, on real-world queries, the amount of self-joins, i.e., when one attribute of a relation needs to be joined to an attribute of that same relation, can be quite important and is known to have poor performances especially on very large relations.

There exists several other approaches to represent RDF Triples in a set of relations, e.g., clustered property, property-class and vertical partitioning [1]. These systems generally accept SPARQL queries which are being translated to SQL queries that the RDBMS understands. The overall performance of query processing is considered not to suffer from these translation steps. Some of the NoSQL stores have also been used to store some RDF Triples. It is for instance the case of Trinity.RDF [20] that uses a key value store on Microsoft's Azure cloud, MarkLogic[10] with a document-based approach and Rya [16] which uses a column-oriented store. A main advantage of these last systems consists of their capacity to be distributed over a large number of machines which makes them particularly adapted to a cloud computing setting. In terms of query

Subject	Property	Object
id1	type	Music
id1	title	"Nefertiti"
id1	author	id5
id1	year	1968
id2	type	Book
id2	title	"The road"
id2	author	"Cormac McCarthy"
id2	year	2006
id2	language	"english"
id3	type	DVD
id3	title	"Into the Wild"
id3	year	2007
id4	type	Music
id4	title	"1984"
id5	type	Man
id5	fullName	"Miles Davis"
id5	yob	1926

Fig. 5. Triple table example

---

[10] http://www.marklogic.com/.

answering, queries are again expressed in SPARQL and are either translated to specific query language accepted by the underlying NoSQL store or to some API (Application Programming Interface) calls.

The systems belonging to the native category do not rely on an existing DBMS. Hence, they generally design their architecture from scratch. Although being time consuming to design and implement, it enables to define a solution that is particularly adapted to graph data model of RDF. In these systems, a peculiar attention is given to the indexing solution, *i.e.*, to ensure high query answering performances, as well as query processing optimization which are dedicated to SPARQL. Examples of native triple store implementations include OWLIM/GraphDB, Virtuoso, Mulgara.

The market currently proposes production-ready RDF stores which are almost evenly distributed over the native and non-native categories.

4.2 SPARQL

There are two versions of the SPARQL specifications: SPARQL 1.0 (2008) only supports information retrieval query facilities and the SPARQL 1.1 (2013) that, in addition to adding new query capabilities, supports data updates. In this subsection, we concentrate solely on the query aspect of SPARQL. Concerning the other functionalities supported by this technology, e.g. service description, one is invited to read the SPARQL 1.1 recommendations[11].

Introduction. The SPARQL query language operates over RDF graphs made of triples and processes queries by adopting a graph pattern matching approach. Intuitively, a SPARQL query specifies a set of so-called graph patterns (GPs) corresponding to triple patterns (TP) composed of a subject, a predicate and an object. A difference compared to the triples encountered in RDF documents is that, in a TP, any of the three positions can be a variable. In that situation, using the same variable over different TPs consists in creating some joins. Hence the principle of SPARQL query answering is to find bindings, i.e. values, for these variables, when executed over a given graph. This approach is denoted as graph pattern matching due to the fact that we are trying to find matches between the set of GPs of a query over an given graph. In the next subsection, we provide an example of graph pattern matching.

The 4 Basic Commands. The syntax of the SPARQL query language resembles the one of SQL. This has been motivated by the popularity of SQL which is present in all RDBMS. Hence the learning curve of SPARQL is fast for users mastering SQL. The main operators of SPARQL are SELECT, ASK, CONSTRUCT, DESCRIBE. In order to retrieve some information from a database, one writes a SELECT query which contains a WHERE clause and possibly a FROM clause, which identifies target graphs. In the SELECT clause, one specifies the variables (s)he

---

[11] http://www.w3.org/TR/sparql11-overview/.

wants to see in the result set. Such variables are denoted as distinguished. In the WHERE clause, we declare the GPs that are supporting the graph pattern matching.

Let us consider the query that retrieves the full name of males that have authored some music document. This is expressed in SPARQL as:

```
SELECT ?fn
WHERE
    {
        ?x type Music.
        ?x author ?y.
        ?y fullname ?fn.
        ?y type Man.
    }
```

This query can be represented as the graph of Fig. 6(a). Executing this query over the data set of Fig. 5 corresponds to finding a subgraph in Fig. 6(b), which represents that data set, and to associate values to the variables of the query. For that query, we have a match for ?x=id1, ?y=id5 and ?fn=''Miles Davis''. Note that only the ?fn will be displayed in the answer set since it is the only variable appearing in the SELECT clause of that query.

Typical SELECT queries can also contain operators such as LIMIT, OFFSET, ORDER BY, DISTINCT that SQL users already know. It also support aggregation with GROUP BY, HAVING and the standard max, min, sum, avg, count functions. Moreover, one can obtain the union of two queries, perform some regular expression operations using the FILTER keyword, execute some outer joins with OPTIONAL.

SPARQL possesses some other operators which are using the notion of GP. ASK queries are boolean queries, i.e., returning true or false, on the existence of

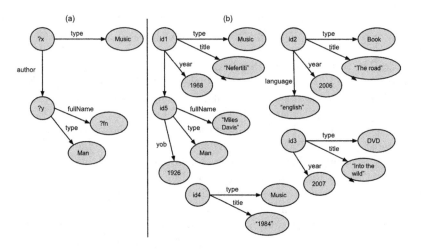

Fig. 6. Graph pattern matching example

a given graph pattern. The form of a CONSTRUCT query is quite similar to the one of SELECT query except that it does not specify any distinguished variable. In fact, the answer to such a query is a new graph. This is quite efficient if one wants to integrate RDF graphs from queries obtained from another graph. Finally, DESCRIBE provides for any RDF information related to some specific variables of the result template. This is a nice approach to learn about what a graph knows about some node.

The last SPARQL recommendation, denoted 1.1, also proposes support for update operations. Compared to SQL, it limits its patterns to DELETE and INSERT, so no UPDATE like in SQL. They respectively correspond to removing and adding some statements from (resp. into) a given RDF graph. We will see in Sect. 6 that in the context of reasoning, the operational aspect of these directives are quite involved.

Limits. RDF Stores correspond more to data management systems that support discovery than to systems supporting high transaction rates. Hence, we can consider that SPARQL is lacking some analytic features. Moreover, support for efficient management of update operations is poor in current RDF Stores.

4.3   SPARQL Endpoint and Federation

As its name implies, the SPARQL specifications define both a query language dedicated to RDF and a protocol for accessing RDF stores. This protocol[12] supports two standard bindings (HTTP and SOAP - a lightweight W3C protocol intended for exchanging structured information in a decentralized, distributed environment) and defines the request parameters as well as the supported response formats.

The SPARQL endpoint is a key feature of RDF stores as it ensures interoperability between all implementations. Hence a change in the RDF store back end will have no impact on the applications accessing the database (provided they do not make use of any provider-specific extensions, such a plain-text search that some products offer).

In addition to the standard bindings, most RDF stores also provide a basic HTML GUI (Graphical User Interface) to their SPARQL endpoint for the users to browse the store contents. Query results are usually displayed as simple HTML tables.

Although SPARQL 1.1 added support for modifying data (create, update and delete), most public endpoints, for obvious security reasons, block write access to the data. Although this can be achieve through access rights managements, some tools (Datalift for example) favor providing a read-only SPARQL endpoint that acts as a reverse proxy, shielding the underlying RDF store from direct access from the Web. In addition to blocking update queries, this front-end can include additional features such as support for the SPARQL 1.1 Graph Store HTTP Protocol[13] (when not provided by the RDF store endpoint) or

---

[12] http://www.w3.org/TR/sparql11-service-description/.
[13] http://www.w3.org/TR/sparql11-http-rdf-update/.

graph-based access control. They may also provide value-added user interfaces, such as the Flint SPARQL Editor,[14] to assist SPARQL query editing (syntax highlighting, auto-completion...) or results browsing.

Scalability of RDF stores is today limited to several billions of triples. As RDF stores are schema-less databases and the SPARQL language puts no restrictions to the complexity of the queries or the scope of data they may apply to, it is quite difficult to efficiently split RDF data into partitions (sharding) and then route a query to only a subset of the partitions.

The SPARQL Federated Query specification[15] helps addressing these shortcomings by defining an extension to the SPARQL query language that allows a query author to direct a portion of a query to a particular SPARQL endpoint. Results are returned to the federated query processor and are combined with results from the rest of the query.

Uses for SPARQL federations are:

- Spread one organization's datasets on several RDF stores if their combined sizes exceed the capacity of the single RDF store installation.
- Perform join queries on a set of RDF datasets hosted on independent RDF stores without having to copy these data locally.

The main drawback of SPARQL Federated Query is that it mandates the query author to know which data are available at each SPARQL endpoint and explicitly craft the query from this knowledge. In the following example, note the SERVICE clause that enables to retrieve some data from a given URI:

```
PREFIX foaf: <http://xmlns.com/foaf/0.1/>
SELECT ?name
WHERE
   {
       <http://example.org/myfoaf/I> foaf:knows ?person .
       SERVICE <http://people.example.org/sparql>
       { ?person foaf:name ?name . }
   }
```

4.4   The Data-Lifting Process

During his talk at the "Gov2.0 Expo" in 2010, Tim Berners-Lee presented a scale to rate open data [7]:

- 1 star: Make your data available on the Web (whatever the format) under an open license
- 2 stars: Make it available as machine-readable structured data (e.g. Excel instead of image scan of a table)
- 3 stars: Use a non-proprietary format (e.g., CSV instead of Excel)

---

[14] http://openuplabs.tso.co.uk/demos/sparqleditor.
[15] http://www.w3.org/TR/sparql11-federated-query/.

– 4 stars: Use URIs to identify things, so that people can point at your stuff
– 5 stars: Link your data to other people's data to provide context.

The Datalift project[16] is one of the (many) initiatives that were launched to provide tool sets to help users publishing 5-star linked open data.

The challenge here is that of a virtuous circle where quantity calls for quality. The more data is linked, and the more it becomes a reference that can attract new data to be connected to. Efforts made by some to link their data will benefit others and have a positive feedback. In this game, every new link takes advantage of those already established, and all existing connections benefit from new contributions.

The raw data from which we have to start are heterogeneous, generally unstructured and stuck at the bottom of information systems. They are generally quite difficult to handle by users. It is possible to refer to them by file name or to extract them from a database but they often do not comply with standards. Their structure is at most syntactical.

As seen above, for the users to make the most out of the published data, it is necessary to explicit the data semantics, by using appropriate vocabularies, also called ontologies. A good practice is to avoid creating a new ontology for any new dataset, but to start with those that already exist. Ontologies durability and reuse rate can instruct us on their efficiency. By efficiency, we must understand the power to make the web less sparse. Indeed, each new usage increases the expressiveness of the data already described, making them more universal.

So, the platform should include features such as ontology management. Among past initiatives, we know that a simple catalog or a search engine is not enough. Ontologies describe knowledge, their management therefore requires discovery heuristics based on existing or deducted links.

The Datalift open-source project aims at providing a one-stop shopping solution to transform raw datasets into RDF structures according to a set of selected ontologies (see Fig. 7). As described above, RDF is a simple yet powerful model for representing knowledge on the web using a triple pattern resulting in a graph-based representation of the data. The transition to RDF makes the resulting data self-expressive. Usage of public and well-known, and widely used ontologies reinforce this expressiveness.

Thus the first step of the data-lifting process is the selection of a (set of) ontologies. Rather than letting users searching for such ontologies, the project maintains an ontology catalog: the Linked Open Vocabularies (LOV)[17]. This catalog only references open RDFS (RDF Schema) and OWL (Web Ontology Language) ontologies, classifies them into business-oriented categories (spaces) and provides a powerful search engine.

Once the ontologies applying the user raw dataset is selected, the data transformation process begins by first converting the data into RDF. This first RDF format derives directly from the raw data format, making use of every possible

---

[16] http://datalift.org/.
[17] http://lov.okfn.org/.

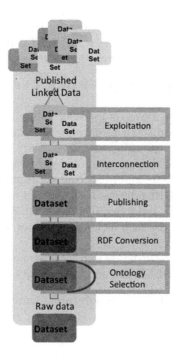

Fig. 7. The data-lifting process

metadata available (column names in CSV files, data types in SQL databases...). Datalift then proposes a set of modules to incrementally transform these basic RDF data into refined ones that match the structure defined by the selected ontologies. These transformation steps may be performed several times as, for complex datasets, each ontology may only model part of the business domain. Each transformation step produces a new RDF graph, a *Named Graph*, which is an administrative object managed by RDF stores to group related RDF triples, allowing to manipulate them as a whole. Until data are ready to be published, these graphs are stored in an internal RDF store.

Once the transformation steps are completed, data can be published, by copying the corresponding named graphs into a public RDF store. Yet, just as data alone has no value at the unit level, and that only the links matter, the same applies at graph level. By publishing the RDF graphs built in the previous steps into a publicly-accessible triple store, we open a new potential of linking: to intra-graph links, it becomes possible to add inter-graph links.

This is the interconnection step: it is possible to find relationships between data, *e.g.*, discovering identical resources (two URI identifying the same real world object) in different graphs, local or remote (*i.e.* hosted on distant SPARQL endpoints). Interconnection can occur either at the dataset level (by comparing RDF resources) or at the ontology level. Ontology alignment (mapping, matching) is an active research topic where new metrics are developed to perform data integration at the schema level rather than at the object level.

The final step is the exploitation of the published linked data. While Datalift directly exposes published RDF resources over HTTP, it also provides additional modules to ease the access and consumption of RDF data:

- A SPARQL endpoint supporting the SPARQL 1.1 syntax but limited to read-only access
- An RDF-based data access control module, S4AC (Social Semantic SPARQL Security for Access Control), that controls which data are accessible to each user: two users running the same SPARQL query will get different results, depending on the graphs they are each allowed to access
- Tools to enforce URI policies and tailor content negotiation. These tools allow, for example, to set up URI policies distinguishing representation URLs (one for each supported MIME type) from canonical RDF resource URIs or provide alternative representations, such as legacy GML or WKT representations for GeoSPARQL data
- Tools to help developing RDF-based web applications, *e.g.* generation of HTML pages or deployment of REST web services relying on SPARQL queries.

Figure 8 depicts the Datalift platform software architecture.

5   Ontologies

Ontologies are at the heart of many applications of knowledge engineering, especially the Semantic Web. They correspond to the expression of some knowledge and thus provide a vocabulary consisting of concepts together with their relationships. They support knowledge management and reasoning facilities and hence offer semantic interoperability between agents, either human or computerized ones. In this section, we begin by addressing the issue of creating ontologies in a mainly methodological angle. Then we will discuss the issues related to their management and we will be referring to many tools. We conclude with a very exciting topic: ontology alignment.

5.1   Ontology Creation

One has to understand that creating an ontology is a time consuming and difficult task. A first step generally consists in searching if an existing ontology can be reused and/or modified. This would make the creation process much easier.

Because an ontology expresses the view of its designer, it could be appropriate if it matches the right knowledge domain, but inappropriate due to a wrong point of view. We could also use portions of existing ontologies to cover part of the problem and create only some complements. The first step is therefore methodological and we can state some good practices.

- Meet those who know: Without domain experts and users of applications, it is not possible to formalize their knowledge and needs, to establish a common vocabulary and to build a consensus (even if it is very hard to read) on the concepts of their domain.

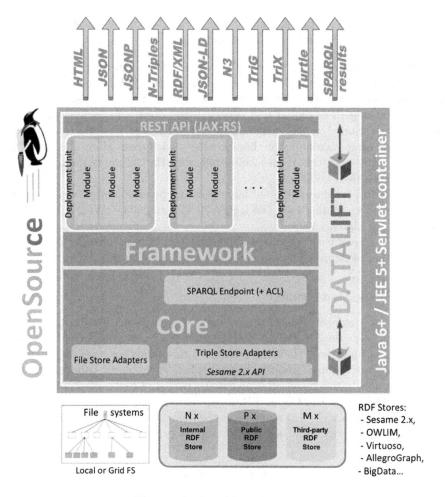

Fig. 8. The Datalift architecture

- Create an ontology according to usage: A good ontology is not the one that meets to the whole domain, but which is guided by usage. What are the questions that the ontology must answer? Although an ontology seems to be a particularly abstract object, it must be limited to the concrete needs we have.
- Do not repeat what has already been done by others: We will take this matter further (see Catalog).
- Do a work of quality and be responsible: When coding the ontology, we must provide care for its expression. The syntax of the language is not too demanding, so we have to impose ourself some rules (mention the language, place a cartridge, comment, etc.). Being responsible is perhaps the least recognized best practice. Yet, it is important to feel responsible for the ontology created according to those who use it. The ontology is a reference, it should evolve and

be maintained as long as it will be used. Other important questions concern who will use and maintain the ontology?

There is no miracle tool to create an ontology. We can adopt a bottom-up approach based on a corpus. It is very common to start by building lexicons from domain materials and semi-automated software that extract terms. We can also analyze an area with an expert and develop a top-down approach. A mind-map tool, *e.g.* FreeMind,[18] can be very useful here. It is also possible to start with weakest structures (taxonomy or thesaurus) to enrich them after. The literature abounds. In any case, this is an intellectual work that requires thought, a step back, assessments ... and time.

To express an ontology, different languages and possible syntaxes can be used, Table 1. The most common are OWL (OWL1 in Feb. 2004, OWL2 in Oct. 2009) and RDFS (in March 1999).

RDFS is a very basic ontology language. It is easy to use and get started with but it is less expressive than the OWL languages (more on this in Sect. 6).

5.2   Ontology Management

Ontologies are complex objects. Some may be very large, for example in the medical field, *e.g.*, the SNOMEDCT (Systematized Nomenclature of Medicine - Clinical Terms) has more than three hundred thousands concepts. So, tools are needed to edit, transform (modify, merge, split), view, query, validate, catalog, etc. We can not, in this pages, give an exhaustive list of tools. We limit ourselves to a short description of LOV and Protégé.

LOV provides a technical platform for search and quality assessment among the vocabularies ecosystem, but it also aims at promoting a sustainable social

Table 1. OWL2 syntaxes, (source : http://www.w3.org/TR/owl2-overview/)

Syntax name	Specification	Status	Purpose
RDF/XML	Mapping to RDF graphs, RDF/XML	Mandatory	Interchange (can be written and read by all compatible OWL 2 software)
OWL/XML	XML serialization	Optional	Easier to process using XML tools
Functional syntax	Structural specification	Optional	Easier to see the formal structure of ontologies
Manchester syntax	Manchester syntax	Optional	Easier to read/write DL ontologies
Turtle	Mapping to RDF graphs, turtle	Optional, not from OWL-WG	Easier to read/write RDF triples

---

[18] http://freemind.sourceforge.net/wiki/index.php/Main_Page.

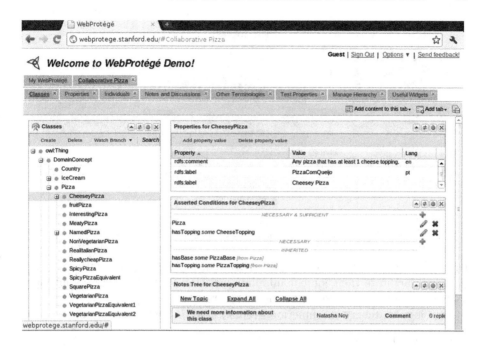

Fig. 9. WebProtégé

management of this ecosystem. Beyond the LOV, there is the vision of a future linked data Web supported by a living Vocabulary Alliance gathering as many as possible stakeholders in the long-term conservation of vocabularies.

Protégé is an ontology editor designed at the Stanford University, distributed as an open-source project and whose catalog of plug-ins[19] provides a good idea of the tools it is possible to find. These plug-ins tackle domains such as visualization, natural language processing, export, import, navigation, query and reasoning. WebProtégé is a Web-based lightweight ontology editor but its goal is not only to be another ontology editor. The purpose is to offer an easier tool and to fill in a significant gap in the landscape of ontology tools. First it is devoted to a large spectrum of users, from ontology experts to domain experts. The user interface (Fig. 9) is customizable by the users with different levels of expertise. It is built like a portal and is composed of predefined or user-defined tabs. The previous version, Protégé, was considered difficult to use.

WebProtégé supports collaboration, ontology reuse and interlinking. It creates an open, extensible and flexible infrastructure that can be easily adapted to the needs of different projects [13].

5.3   Ontology Alignment

Ontology matching [12] is the process used to establish semantic relations between concepts and relations of two ontologies. If the ontologies are designed to

---

[19] http://protegewiki.stanford.edu/wiki/Protege_Plugin_Library.

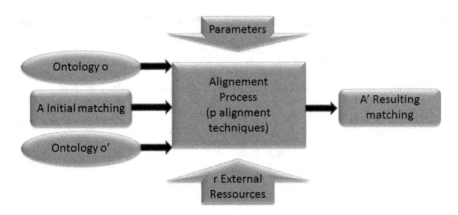

Fig. 10. The matching process, A' = f(o; o'; A; p; r)

neighboring areas, with analogies, similarities or any form of correlations, alignment will facilitate their joint manipulation within a single system or multiple
systems using these ontologies to communicate.

Ontology matching is a function f represented in Fig. 10 and applied:

– to two ontologies o and o',
– to a set of original matching A,
– from a set of parameters applicable to p alignment techniques implemented
  in the process
– and a set of external resources r,
– and that produces a set of matching A' between the two ontologies.

Alignments express correspondences between entities belonging to different
ontologies.

6   Reasoning in a Semantic Web Context

6.1   Introduction

In this section, we present the role played by ontologies in the process of deriving
novel information from a set of RDF statements. We then introduce the different
methods on which reasoners have based their implementation.

The reasoning aspect is reminiscent of the Knowledge Representation (KR)
field of Artificial Intelligence (AI). Reasoning can be described as the ability to
deduce implicit knowledge from explicitly one. Here, we consider that the concept
of knowledge subsumes the previously introduced notions of information, data
and obviously knowledge. In the context of the Semantic Web, this implies a
Knowledge Base composed of a set of RDF facts together with a non empty set
of ontologies expressed in RDFS and/or OWL. Note that such derivations can
not be performed in the absence of at least one ontology since it is the component
responsible for storing the knowledge. On the other hand, it is possible to reason
over an ontology alone, i.e., without facts.

This is clarified with a simple example where the fact base is limited to the assertion father(Camille, Pierre) and the ontology comprises some concept descriptions, Man and Woman are disjoint subconcepts of Person, and two property descriptions, father and parent both have the Person concept as their range and the Man, resp. Person, concepts as domain. First, note that to represent this ontology, the expressive power of RDFS is not sufficient. This is due to the disjointness between the Man and Woman concepts which requires a negation. Since the Fact base does not provide any description on either concept and property, it is not possible to infer anything interesting from this Knowledge base component. With the ontology alone, we can deduce that a Man or Woman instance can be involved in parent relationship, either as the parent or the child. Finally, with both the ontology and fact base, we can infer that Camille and Pierre are respectively instances of the Man and Person concepts.

Thus, ontologies are supporting the ability to reason over the facts and depending on their expressive power different inferences can be obtained. The expressive power of an ontology is essentially associated to the constructors provided by the ontology language, i.e., concept negation, disjunction, quantifiers, cardinalities to name a few. Obviously, the more expressive the ontology language, the more inferences can be deduced but this also comes at a certain cost. Previous work in Description Logics (DL) [5] have emphasized a trade-off between the expressive power of an ontology and the computational complexity. To simplify, the more expressive the ontology language, the longer the duration, in favorable cases where the processing terminates, one has to wait to obtain the correct complete set of deductions. For very large Knowledge Bases, this duration may correspond to hours, days or more. By itself, this trade-off motivates the existence of the different ontology languages of the Semantic Web with RDFS being the less expressive and OWL2DL being the most expressive among the decidable ones, i.e., those ensuring that the deduction process will terminate. For instance, OWL2 Full is known to be undecidable. In between RDFS and OWL2DL lies some languages like the OWL Lite and the more recent OWL2QL, OWL2EL and OWL2RL which have precisely been designed to provide interesting expressiveness/computational complexity compromises. Figure 11 presents the expressiveness containment relationships of RDFS and OWL2 ontology languages.

So far, we have not said anything on the "how" these reasoning services are being performed over a KB. Conceptually, two different approaches can be used to address the reasoning task.

6.2 Procedural Approach

With a procedural approach, the inferences are hand coded in some programming language, e.g., Java programming language. Although it can reply to some special domain dependent needs, this approach is generally associated with the following drawbacks:

- adding some new reasoning functionalities for an existing system requires programming skills and a very good understanding of the programs. One can not expect a non-programmer to perform this task, e.g., a biology expert will not learn the Java to add some new inference services.

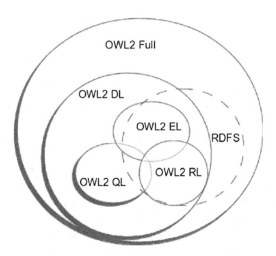

Fig. 11. Expressiveness of RDFS and OWL2 ontology languages

– reusing the reasoning services from one application domain to another involves too much effort from both the programmers and the domain experts. This would require the ability to identify and extract some reasoning patterns encoded in the syntax of a programming language and modify them for another application domain. This is definitely not a task that can be performed automatically.
– enabling the system to explain its inferences, a feature expected from a KB system, is very complex. This would again imply a form of understanding of the intention of the code lines of the program. A thing that we already seen in the two previous bullet points is not possible.

6.3   Declarative Approach

The second approach is characterized as declarative. Intuitively, an inference engine performs some tasks on a external component that is responsible for storing the facts and can communicate with an application. This separation between these three components is a differentiator to the previous procedural approach where the inference engine, a.k.a. a reasoner, and the application were kind of merged. This separation eases the maintenance of both the reasoner and the application due to a clear separation between the responsibilities of each component. The reasoner is a generic, *domain independent*, logic-based, i.e., adapted to a specific ontology language, and optimized implementation that is able to make some deductions. These two characteristics are supported by operating directly on the logic axioms of the knowledge base. Hence, the same reasoner can be used on any domain, e.g., biology, finance, culture, as long as a compliant KB is provided. It also enables any domain expert to maintain the set of axioms using a user-friendly, graphical interface such as the Protégé KB editor. Moreover, these reasoners are generally equipped with components that enable to explain inferences, usually by going through the different reasoning steps. Due to its nice properties, the declarative approach is more frequently used than the procedural one.

6.4   Declarative Reasoning Methods

There exists several methods to perform inferences within the declarative approach. They can be classified into two categories depending on the representation of the knowledge, that is either rule or object based.

In the former, the inference is performed using the resolution principle which has been introduced by Robinson in [17]. It was used in the area of automated theorem-proving by trying to derive a theorem from a set of axioms and proceeds by construction a refutation proof. It is easily adapted to standard ontology languages such as RDFS, the RL and EL OWL2 fragments, to some extent to OWL2 QL as well as the non-standard RDFS+ and OWL Horst languages.

In the object-based category the main inference pattern is based on the notion of inheritance, i.e., subsumption relationships between concepts or properties. For instance, it is particularly adapted to compute the concept hierarchy in an OWL ontology. Depending on the expressiveness of the underlying ontology, it is most frequently computed with a structural subsumption algorithm or a tableau-based one (but automata-based approaches or translation to a First Order logic reasoner are possible). In the former, the inheritance is strictly computed at the syntax level of the concept descriptions. That is one searches if the concept, denoted as C, corresponding to a happy man whose parents are all doctors is a specialization, i.e., a subconcept, of the concept, denoted as D, described as a man whose parents are all doctors. The problem can be stated as "is $C \sqsubseteq D$?" and intuitively tries to find if all elements on the D definition has an equivalence in the C description. If this is the case then the subsumption between these two concepts holds otherwise it does not. For instance, the concept *Man* and *all parents are doctors* are on both sides of the $\sqsubseteq$ symbol. So we remove these concepts on both sides, thus nothing remains on the right side (D) and the concept *Happy* remains on the left side C. So it is the case that $C \sqsubseteq D$. This method is adapted to ontologies with a low level of expressiveness, e.g., OWL2 EL, since for more expressive ones, it may not be able to perform the task appropriately. The tableau-based approach is a widely used approach and is the one we encounter in well-known systems such as Pellet, RacerPro, Fact++ and to some extent HermiT (which is now integrated within the Protégé editor). Just like in the resolution approach, the method relies on proof by contradiction using a set of so-called expansion rules. Additionally to the classification service, the tableau method can perform other standard services such as consistency checking, i.e., detect whether a KB is consistent or not, and so-called realization and retrieval which respectively find the most concept an object is an instance of and find the individuals that are instances of a given concept. Some reasoners even support some denoted non-standard services and provide some explanations on their inferences. This approach is particularly adapted to expressive ontology languages such as the decidable OWL DL and OWL2 DL.

It is important to stress that the SPARQL query language does not support any reasoning service. Hence, the kind of answer it returns for a query do not natively contain any inferred statements. In our running example, asking for instances of the `Man` concept would return an empty set.

In order to reply complete results to query requiring inferences, two approaches are used by RDF stores. In the first one, the complete set of facts that can inferred are stored in the database. This materialization is performed when the data set is first loaded in the RDF store or where the database is updated. Consider the KB of our previous example, one of the materialized statement would state that `Camille` is of type `Man`, represented as `Man(Camille)`. Thus, this approach is quite efficient in terms of query answering performance since no additional inferences are performed at query run-time. That is `Camille` would be part of the query asking for man instances. Nevertheless, it comes with the following limitations: the size of stored database can grow exponentially compared to the size of the original data set, the loading time of the data set can be quite important and handling updates is a complex task since one basically needs to know if a statement has been inferred or is part of the original data set. For instance, consider that we modify the KB of our running example by substituting `Camille` by `Joe` in the `father(Camille, Pierre)` assertion. Then the system would have to remove the `Man(Camille)` from the database since that fact does not hold anymore.

The second solution to support deductions in RDF stores is based on the introduction of a reasoning step at query run-time. This implies to rewrite the original query into a possibly large number of new queries that are all going to be executed. The final answer set then consists of the union of the result set of each query. In our running example, the query asking for all man occurrences would result in the execution of the original query plus the query asking for the domain of `father` property assertions. Obviously, the correct `Camille` result would be obtained from that last query. Of course, the integration of the reasoning step has the negative impact of slowing down query answering. Nevertheless, the size of the stored data is not expanded and updating the database is less demanding than in the materialization solution.

7   Research and Development Perspectives

7.1   Introduction

As emphasized in previous sections, available Web of data building blocks are covering expected functionalities and are of sufficient quality to be used in production. It is for instance the case for the RDF data model, the spectrum of ontology languages with a W3C recommendation status and their associated reasoners as well as the SPARQL query language. This permits to address data integration in an efficient way, an aspect that is particularly interesting in the context of Linked Open Data. During the last few years, we have witnessed an acceleration of the adoption of these technologies in the information technology industry. For instance, more and more web sites are annotating their content with Schema.org. A blog entry at SemanticWeb.com[20] announced that in September 2014, 21 % of a sample of 12 billions web pages were using that solution for annotation purposes. Moreover it is said that "every major site in every major category, from news to e-commerce (with the exception of Amazon.com), uses it." As a last example, the Knowledge Vault project [11] developed at Google and which is believed to replace the Knowledge Graph for improving semantic search, claims to use RDF Triples to represent extract information.

We nevertheless consider that more is needed to speed up the adoption of these technologies at a large scale. In the remaining of this section, we present some directions that we believe are primordial to attain the vision of the Web of data and of the Semantic Web.

7.2 Big Data

We can first wonder what is the stance occupied but the technologies of the Web of data in the Big Data phenomenon. Although a large set of RDF Triple stores are being produced, we consider that a robust, equipped with efficient support for updates operations and reasoning services, is needed. The current state of RDF stores does not address these issues with sufficient guarantees. For instance, we know of several large projects that are using some commercial, state of the art RDF stores but which are still not using their update operations or reasoning functionalities due to poor performances. That is these companies prefer to use a bulk loading approach, rather than an incremental one, when new data are to be integrated in the RDF Store. Intuitively, this means that updates are kept somewhere, waiting to be sent to the store together with the previous data. The main drawback is the staleness of the database. Moreover they are mostly not using any forms of inferences although their use cases would probably benefit from it. The data and processing distribution aspects are relatively well managed by existing commercial and production ready RDF Stores. These database systems are mainly used for data integration and analytics purposes rather than to process high rates of transactions, due to their relatively poor update performances especially in the situation of a materialization approach. Thus these systems are more used as data warehouses, relevant to the OnLine Analytical Processing (OLAP), rather than the OnLine Transactional Processing (OLTP) movement. Within this context, we can argue that the SPARQL query language is missing some analytical operations toward the management of aggregations and dimensions, e.g., drilling down and rolling up, slicing and dicing. We believe that these tools should benefit from the concept and property hierarchies already represented in ontologies. Other operations such as graph navigation features are needed to compete with graph store of the NoSQL ecosystem. The integration of statistical languages, such as the R language, would also open some new perspectives to end-users in need to take some decisions. In that same data analytic aspect, libraries and tools for data visualization and user interfaces are needed by data scientists that are manipulating RDF data.

---

[20] http://semanticweb.com/schema-org-fires-lit_b44380.

7.3 Data Streams

Still related to the data deluge aspects, it seems that methods to handle streams in both the processing and reasoning aspects are highly expected from the community of developers and designers of Web of data applications. An important number of researchers in the field of database management systems is currently working on this subject which addresses the velocity characteristic of Big Data. In particular, [18] presents a set of requirements expected from such systems in the context of RDBMS. Considering streams in the context of the Web of data implies to tackle the generation or efficient transformations of data emanating from different kind of sensors, the capacity to analyze, filter, store, summarize them as well as reason over them or their summarizations. These are aspects that have started to be investigated for the RDF data format, for instance with the proposition of SPARQL extensions [6,14].

7.4 Linked Open Data

Other expectations concern Linked Open Data and in particular the quantity of linked data sets and the quality of the proposed links[21]. We consider that these two notions are related. It is clear that the more linked open data sets, the more mash-up application can be designed. Nevertheless, a notion of quality has to be considered and recent studies [10] highlighted that the owl:sameAs property is widely used with different meaning. Just like in recent automatic KB construction solutions [11] a probabilistic approach should be used to qualify the confidence on the link between entities present in different ontologies or KB.

7.5 Other Aspects

Finally and not least, it is obvious that security, privacy which is related to identity management in the virtual world of the Web will play an important role in fulfilling the vision of the Semantic Web.

References

1. Abadi, D.J., Marcus, A., Madden, S., Hollenbach, K.J.: Scalable semantic web data management using vertical partitioning. In: Proceedings of the 33rd International Conference on Very Large Data Bases, University of Vienna, Austria, 23–27 September 2007, pp. 411–422 (2007)
2. Abiteboul, S., Hull, R., Vianu, V. (eds.): Foundations of Databases: The Logical Level, 1st edn. Addison-Wesley Longman Publishing Co., Inc., Boston (1995)
3. Ackoff, R.: From data to wisdom–presidential address to ISGSR.J. Appl. Syst. Anal. **16**, 3–9 (1989)
4. Amidon, D.M.: Innovation Strategy for the Knowledge Economy: The Ken Awakening. Routledge, London (1997)

---

[21] http://lod-cloud.net/.

5. Baader, F., Calvanese, D., McGuinness, D.L., Nardi, D., Patel-Schneider, P.F. (eds.): The Description Logic Handbook: Theory, Implementation, and Applications. Cambridge University Press, New York (2003)

6. Barbieri, D.F., Braga, D., Ceri, S., Della Valle, E., Grossniklaus, M.: C-sparql: sparql for continuous querying. In: Proceedings of the 18th International Conference on World Wide Web, WWW 2009, pp. 1061–1062. ACM, New York (2009)

7. Berners-Lee, T.: Linked data-design issues (2006). http://www.w3.org/DesignIssues/LinkedData.html (2011)

8. Cotton, F.: Politique d'identification des ressources (2012). https://gforge.inria.fr/docman/view.php/2935/8286/D3.1+Politique+dpdf

9. Curé, O., Guillaume, B. (eds.): RDF Database Systems: Triples Storage and SPARQL Query Processing, 1st edn. Morgan Kaufmann, Boston (2015)

10. Ding, L., Shinavier, J., Shangguan, Z., McGuinness, D.L.: SameAs networks and beyond: analyzing deployment status and implications of owl:sameAs in linked data. In: Patel-Schneider, P.F., Pan, Y., Hitzler, P., Mika, P., Zhang, L., Pan, J.Z., Horrocks, I., Glimm, B. (eds.) ISWC 2010, Part I. LNCS, vol. 6496, pp. 145–160. Springer, Heidelberg (2010)

11. Dong, X., Gabrilovich, E., Heitz, G., Horn, W., Lao, N., Murphy, K., Strohmann, T., Sun, S., Zhang, W.: Knowledge vault: a web-scale approach to probabilistic knowledge fusion. In: The 20th ACM SIGKDD International Conference on Knowledge Discovery and Data Mining, KDD 2014. New York, NY, USA, pp. 601–610, 24–27 August 2014

12. Euzenat, J., Shvaiko, P.: Ontology Matching, 2nd edn. Springer, Heidelberg (2013)

13. Horridge, M., Tudorache, T., Vendetti, J., Nyulas, C.I., Musen, M.A., Noy, N.F.: Simplified OWL ontology editing for the web: Is webProtégé enough? In: Alani, H., Kagal, L., Fokoue, A., Groth, P., Biemann, C., Parreira, J.X., Aroyo, L., Noy, N., Welty, C., Janowicz, K. (eds.) ISWC 2013, Part I. LNCS, vol. 8218, pp. 200–215. Springer, Heidelberg (2013)

14. Le-Phuoc, D., Dao-Tran, M., Xavier Parreira, J., Hauswirth, M.: A native and adaptive approach for unified processing of linked streams and linked data. In: Aroyo, L., Welty, C., Alani, H., Taylor, J., Bernstein, A., Kagal, L., Noy, N., Blomqvist, E. (eds.) ISWC 2011, Part I. LNCS, vol. 7031, pp. 370–388. Springer, Heidelberg (2011)

15. Liew, A.: Understanding data, information, knowledge and their interrelationships. J. Knowl. Manage. Pract. **8**(2), (2007)

16. Punnoose, R., Crainiceanu, A., Rapp, D.: Rya: a scalable rdf triple store for the clouds. In: Proceedings of the 1st International Workshop on Cloud Intelligence, Cloud-I 2012, pp. 4:1–4:8. ACM, New York (2012)

17. Robinson, J.A.: A machine-oriented logic based on the resolution principle. J. ACM **12**(1), 23–41 (1965)

18. Stonebraker, M., Çetintemel, U., Zdonik, S.B.: The 8 requirements of real-time stream processing. SIGMOD Record **34**(4), 42–47 (2005)

19. Tennison, J.: Creating uris (2010). http://data.gov.uk/resources/uris

20. Zeng, K., Yang, J., Wang, H., Shao, B., Wang, Z.: A distributed graph engine for web scale RDF data. In: Proceedings of the 39th International Conference on Very Large Data Bases, PVLDB 2013, pp. 265–276. VLDB Endowment (2013)

A Survey on Supervised Classification on Data Streams

Vincent Lemaire[1]([⊠]), Christophe Salperwyck[2], and Alexis Bondu[2]

[1] Orange Labs, 2 Avenue Pierre Marzin, 22300 Lannion, France
vincent.lemaire@orange.com
[2] EDF R&D, 1 Avenue du Général de Gaulle, 92140 Clamart, France
{christophe.salperwyck,alexis.bondu}@edf.fr

Abstract. The last ten years were prolific in the statistical learning and data mining field and it is now easy to find learning algorithms which are fast and automatic. Historically a strong hypothesis was that all examples were available or can be loaded into memory so that learning algorithms can use them straight away. But recently new use cases generating lots of data came up as for example: monitoring of telecommunication network, user modeling in dynamic social network, web mining, etc. The volume of data increases rapidly and it is now necessary to use incremental learning algorithms on data streams. This article presents the main approaches of incremental supervised classification available in the literature. It aims to give basic knowledge to a reader novice in this subject.

1 Introduction

Significant improvements have been achieved in the fields of machine learning and data mining during the last decade. A large range of real problems and a large amount of data can be processed [1,2]. However, these steps forward improve the existing approaches while remaining in the same framework: the learning algorithms are still centralized and process a finite dataset which must be stored in the central memory. Nowadays, the amount of 'ble data drastically increases in numerous application areas. The standard machine learning framework becomes limited due to the exponential increase of the available data, which is much faster than the increase of the processing capabilities.

Processing data as a stream constitutes an alternative way of dealing with large amounts of data. New application areas are currently emerging which involve data streams [3,4] such as: (i) the management of the telecommunication networks; (ii) the detection and the tracking of user communities within social networks; (iii) the preventive maintenance of industrial facilities based on sensor measurements, etc. The main challenge is to design new algorithms able to scale on this amount of data. Two main topics of research can be identified: (i) the parallel processing consists in dividing an algorithm into independent queues in order to reduce the computing time; (ii) the incremental processing consists in implementing one-pass algorithms which update the solution after processing

© Springer International Publishing Switzerland 2015
E. Zimányi and R.-D. Kutsche (Eds.): eBISS 2014, LNBIP 205, pp. 88–125, 2015.
DOI: 10.1007/978-3-319-17551-5_4

each piece of data. This paper focuses on the second point: incremental learning applied on data streams.

Section 2.1 defines the required concepts and notations for understanding this article. The incremental learning is formally defined in this section and compared with other types of learning. Section 3 presents the main incremental learning algorithms classified by type of classifiers. Algorithms for data streams are then presented in Sect. 4. Such approaches adapt the incremental learning algorithms to the processing of data streams. In this case, the processed data is emitted within a data stream without controlling the frequency and the order of the emitted data. This paradigm allows to process a huge amount of data which could not be stored in the central memory, or even on the hard drive. This Section presents the main on-line learning algorithms classified by type of classifiers. Section 5 addresses the issue of processing non-stationary data streams, also called "concept drift". Finally, Sect. 6 is dedicated to the evaluation of such approaches: the main criteria and metrics able to compare and evaluate algorithms for data streams are presented.

2   Preamble

2.1   Assumptions and Constraints

In terms of machine learning, several types of assumptions and constraints can be identified which are related to the data and to the type of concept to be learned. This section describes some of these assumptions and constraints in the context of incremental learning in order to understand the approaches that will be presented in Sects. 3 and 4 of this article.

In the remainder of this article, we focus on binary classification problems. The term "example" designates an observation x. The space of all possible examples is denoted by $\mathcal{X}$. Each example is described by a set of attributes. The examples are assumed to be i.i.d, namely independent and randomly chosen within a probability distribution denoted by $\mathcal{D}_{\mathcal{X}}$. Each example is provided with a label $y \in Y$. The term "model" designates a classifier (f) which aims to predict the label that must be assigned to any example x ($f : \mathcal{X} \to \mathcal{Y}$).

Availability of Learning Examples: In the context of this article, the models are learned from representative examples coming from a classification problem. In practice the availability and the access to these examples can vary: all in a database, all in memory, partially in memory, one by one in a stream, etc. Several types of algorithms for different types of availability of the examples exist in the literature.

The simplest case is to learn a model from a quantity of representative examples which can be loaded into the central memory and which can be immediately processed. In other cases, the amount of examples is too large to be loaded into the central memory. Thus, specific algorithms must be designed to learn a model in such conditions. A first way of avoiding the exhaustive storage of the examples into the central memory is to divide the data into small parts *(also called chunks)* which are processed one after the others. In this case, parallelization techniques can be used in order to speed up the learning algorithm [5].

In the worst case, the amount of data is too large to be stored. Data are continuously emitted within a data stream. According to the "data streams" paradigm, the examples can be seen only once and in their order of arrival *(without storage)*. In this case, the learning algorithm must be incremental and needs to have a very low latency due to the potentially high emission rate of the data stream.

Availability of the Model: Data mining is a two-step process: (1) learn the model, (2) deploy the model to predict on new data. In the case of regular batch learning, these two steps are carried out one after the other. But in the case of an incremental algorithm, the learning stage is triggered as soon as a new example arrives. In the cases of time-critical applications, the learning step requires a low latency algorithm *(i.e. with a low time complexity)*. Others kinds of time constraints can be considered. For instance, in [6] the authors define the concept of "anytime" algorithm which is able to be stopped at any time and provide a prediction. The quality of the outcome model is expected to be better and better, while the algorithm is not stopped. The on-line classifier can be used to make predictions while it is learning.

Concept to be Learned: Let us consider a supervised classification problem where the learning algorithm observes a sequence of examples with labels. The class value of the examples follows a probability distribution denoted by P_y. The concept to be learned is the joint probability $P(x_i, y_i) = P(x_i)P(y_i|x_i)$, with x_i a given example and y_i a given class value.

The concept to be learned is not always constant over time, sometimes it changes, this phenomenon is called "concept drift" [7]. Gama [8] identifies two categories of drift: either it is progressive and it is called "concept drift", or it is abrupt and it is called "concept shift". These two types of drift correspond to a change in the conditional distribution of the target values $P(Y|X)$ over time. The algorithms of the literature can be classified by their ability (or not) to support concept drifts.

The distribution of data $P(X)$ may vary over time without changing $P(Y|X)$, this phenomenon is called covariate shift [9]. The covariate shift also occurs when a selection of non-iid examples is carried such as a selection of artificially balanced training examples *(which are not balanced in the test set)* or in the case of active learning [10]. There is a debate on the notion of covariate shift, the underlying distribution of examples (D_X) can be assumed to be constant and that it is only the distribution of the observed examples which changes over time.

In the remainder of this article, in particular in Sect. 5, we will mainly focus on the drift on $P(Y|X)$, while the distribution P(X) is supposed to be constant within a particular context. The interested reader can refer to [9,11] to find elements on the covariate shift problem. Gama introduced the notion of "context" [8] defined as a set of examples on which there is no concept drift. A data stream can be considered as a sequence of contexts. The processing of a stream consists in detecting concept drifts and/or simultaneously working with multiple contexts (see Sect. 5).

Questions to Consider: The paragraphs above indicate the main questions to consider before implementing a classifier:

- Can examples be stored in the central memory?
- What is the availability of the training examples? all available? within a data stream? visible once?
- Is the concept stationary?
- What are the time constraints for the learning algorithm?

The answers to these questions allow one to select the most appropriate algorithms. We must determine whether an incremental algorithm, or an algorithm specifically designed for data streams, is required.

2.2   Different Kinds of Learning Algorithms

This section defines several types of learning algorithms regarding the possible time constraints and the availability of examples.

Batch Mode Learning Algorithms: The batch mode consists in learning a model from a representative dataset which is fully available at the beginning of the learning stage. This kind of algorithm can process relatively small amounts of data (up to several GB). Beyond this limit, multiple access/reading/processing might be required and become prohibitive. It becomes difficult to achieve learning within a few hours or days. This type of algorithm shows its limits when (i) the data cannot be fully loaded into the central memory or continuously arrives as a stream; (ii) the time complexity of the learning algorithm is higher than a quasi-linear complexity. The incremental learning is often a viable alternative to this kind of problem.

Incremental Learning: The incremental learning consists in receiving and integrating new examples without the need to perform a full learning phase from scratch. A learning algorithm is incremental if for any examples $x_1, ..., x_n$ it is able to generate the hypotheses $f_1, ..., f_n$ such that f_{i+1} depends only on f_i and x_i the current example. The notion of "current example" can be extended to a summary of the latest processed examples. This summary is exploited by the learning algorithm instead of the original training examples. The incremental algorithms must learn from data much faster than the batch learning algorithms. Such algorithms must have a very low time complexity. To reach this objective, most of the incremental algorithms read the examples just once so that they can efficiently process large amounts of data.

Online Learning: The main difference between the incremental and the online learning is that the examples continuously arrive from a data stream. The classifier is expected to learn a new hypothesis by integrating the new examples with a very low latency. The requirements in terms of time complexity are stronger than for the incremental learning. Ideally, an online classifier implemented on a data stream must have a constant time complexity ($O(1)$). The objective is to learn and predict at least as fast as the data arrives from the stream. These algorithms must be implemented by taking into account that the available RAM is limited and constant over time. Furthermore, concept drift must be managed by these algorithms.

Anytime Learning: An anytime learning algorithm is able to maximize the quality of the learned model (with respect to a particular evaluation criterion) until an interruption (which may be the arrival of a new example). The algorithms by contract are relatively close to the anytime algorithms. In [12], the authors propose an algorithm which takes into account the available resources (time/CPU/RAM).

2.3 Discussion

From the previous points above the reader may understand there are 3 cases:

- the simplest case is to learn a model from a quantity of representative examples which can be loaded into the central memory and which can be immediately processed;
- when the amount of examples is too important to be loaded into the central memory, the first possibility is to parallelize if the data can be split in independent chunks or to have algorithm able to deal the data in one pass;
- the last case, which is dealt in this paper, concerns the data that are continuously emitted. Two subcases have to take into account if the data stream is stationary or not.

3 Incremental Learning

3.1 Introduction

This section focuses on incremental learning approaches. This kind of algorithms must satisfy the following properties:

- read examples just once;
- produce a model similar to the one that would have been generated by a batch algorithm.

Data stream algorithms satisfy additional constraints which are detailed in Sect. 4.1. In the literature, many learning algorithms dedicated to the problem of supervised classification exist such as the support vector machine (SVM), the neural networks, the k-nearest neighbors (KNN), the decision trees, the logistic regression, etc. The choice of the algorithm highly depends on the task and the need to interpret the model. This article does not describe all the existing approaches since this would take too much place. The following section focuses on the most widely used algorithms in incremental learning.

3.2 Decision Trees

A decision tree [13,14] is a classifier constructed as an acyclic graph. The end of each branch of the tree is a "leaf" which provides the result obtained depending on the decisions taken from the root of the tree to the leaf. Each intermediate node in the tree contains a test on a particular attribute that distributes data in the different sub-trees. During the learning phase of a decision tree, a purity criterion such as the entropy (used in C4.5) or the Gini criterion (used in CART) is exploited to transform a leaf into a node by selecting an attribute and a cut value. The objective is to identify groups of examples as homogeneous as possible with respect to the target variable [15]. During the test phase, a new example to be classified "goes down" within the tree from the root to a single leaf. His path is determined by the values of its attributes. The example is then assigned to the majority class of the leaf, with a score corresponding to the proportion of training examples in the leaf that belong to this class. Decision trees have the following advantages: (i) a good interpretability of the model, (ii) the ability to find the discriminating variables within a large amount of data. The most commonly used algorithms are ID3, C4.5, CART but they are not incremental.

Incremental versions of decision trees have emerged in the 80s. ID4 is proposed in [16] and ID5R in [17]. Both approaches are based on ID3 and propose an incremental learning algorithm. The algorithm ID5R ensures to build a decision tree similar to ID3, while ID4 may not converge or may have a poor prediction in some cases. More recently, Utgoff [18] proposes the algorithm ITI which maintains statistics within each leaf in order to reconstruct the tree with the arrival of new examples. The good interpretability of the decision trees and their ability to predict with a low latency make them a suitable choice for processing large amounts of data. However, decision trees are not adapted to manage concept drift. In this case, significant parts of the decision tree must be pruned and relearned.

3.3 Support Vector Machines

The Support Vector Machines (SVM) has been proposed by Vapnik in 1963. The first publications based on this classification method appeared only in the 90s [19,20]. The main idea is to maximize the distance between the separating hyperplane and the closest training examples.

Incremental versions of the SVM have been proposed, among these, [21] proposes to partition the data and identifies four incremental learning techniques:

- ED - Error Driven: when new examples arrive, misclassified ones are exploited to update the SVM.

- FP - Fixed Partition: a learning stage is independently performed on each partition and the resulting support vectors are aggregated [22].
- EM - Exceeding-Margin: when new examples arrive, those which are located in the margin of the SVM are retained. The SVM is updated when enough examples are collected.
- EM+E - Exceeding-margin+errors: use "ED" and "EM", the examples which are located in the margin and which are misclassified are exploited to update the SVM.

Reference [23] proposes a Proximal SVM (PSVM) which considers the decision boundary as a space *(i.e. a collection of hyperplanes)*. The support vectors and some examples which are located close to the decision boundary are stored. This set of maintained examples change over time, some old examples are removed and others new examples are added which makes the SVM incremental.

The algorithm LASVM [24–26] was proposed recently. This online approach incrementally selects a set of examples from which the SVM is learned. In practice LASVM gives very satisfactory results. This learning algorithm can be interrupted at any time with a finalization step which removes the obsolete support vectors. This step is iteratively carried out during the online learning which allows the algorithm to provide a result close to the optimal solution. The amount of RAM used can be parameterized in order to set the compromise between the amount of used memory and the computing time (i.e. the latency). A comparative study of different incremental SVM algorithms is available in [21].

3.4   Rule-Based Systems

The rule-based systems are widely used to store and handle knowledge in order to provide a decision support. Typically, such a system is based on rules which come from a particular field of knowledge (ex: medicine, chemistry, biology, etc.) and is used to make inferences or choices. For instance, such a system can help a doctor to find the correct diagnosis based on a set of symptoms. The knowledge (i.e. the set of rules) can be provided by a human expert of the domain, or it can be learned from a dataset. In the case of automatically extracted knowledge, some criteria exist which are used to evaluate the quality of the learned rules. The interested readers can refer to [27] which provides a complete review of these criteria.

Several rule-based algorithms were adapted to the incremental learning:

- STAGGER [28] is the first rule-based system able to manage concept drift. This approach is based on two processes: the first one adjusts the weights of the attributes of the rules and the second one is able to add attributes to existing rules.
- FLORA, FLORA3 [29] are based on a temporal windowing and are able to manage a collection of contexts which can be disabled or enabled over time.
- AQ-PM [30] keeps only the examples which are located close to the decision boundaries of the rules. This approach is able to forget the old examples in order to learn new concepts.

3.5 Naive Bayes Classifier

This section focuses on the naive Bayes classifier [31] which assumes that the explanatory variables are independent conditionally to the target variable. This assumption drastically reduces the training time complexity of this approach which remains competitive on numerous application areas. The performance of the naive Bayes classifier depends on the quality of the estimate of the univariate conditional distributions and on an efficient selection of the informative explicative variables.

The main advantage of the approach is its low training and prediction time complexity and its low variance. As shown in [32], the naive Bayes classifier is highly competitive when few learning examples are available. This interesting property contributes to the fact that the naive Bayes classifier is widely used and combined with other learning algorithms, as decision trees in [33] (NBTree). Furthermore, the naive Bayes approach is intrinsically incremental and can be easily updated online. To do so, it is sufficient to update some counts in order to estimate the univariate conditional probabilities. These probabilities are based on an estimate of the densities; this problem can be viewed as the incremental estimation of the conditional densities [34]. Two density estimation approaches are proposed in [35].

IFFD is proposed in [36] which is an incremental discretization approach. This approach allows updating the discretization at the arrival of each new example without computing the discretization from scratch. The conditional probabilities are incrementally maintained in memory. The IFFD algorithm is only based on two possible operations: add a value to an interval, or split an interval. Others non-naive Bayesian approaches exist which are able to aggregate variables and/or values.

3.6 KNN - Lazy Learning

The lazy learning approaches [37] such as the k-nearest neighbors (KNN) do not exploit learning algorithms but only keep a set of useful examples. No global model is built. The predictions are carried out using a local model (built on the fly) which exploits only a subset of the available examples. This kind of approaches can be easily adapted to the incremental learning by updating the set of the stored examples. Some approaches avoid using the old examples during this update [38]. For instance, the KNN approach predicts the label of an example by selecting the closest training examples and by exploiting the class values of these selected examples.

Two main topics of research exists in literature: (i) the speed up of the search of the KNN [39,40]; (ii) the learning of metrics [41–43]. A comparative study of incremental and non-incremental KNN can be found in [44].

4 Incremental Learning on Streams

4.1 Introduction

Since 2000, the volumetry of data to process has increased exponentially due to the internet and most recently to social networks. A part of these data arrives continuously and are visible only once: these kinds of data are called data streams. Specific algorithms dedicated to data streams are needed to efficiently build classifiers. This section presents the main methods in the literature for classification on data streams.

First, let us focus on the expected properties for a good algorithm designed for incremental classification on streams. Domingos wrote a general article [45] about learning on data streams and identify additional expected properties of this kind of algorithms:

- process each example in a low and constant time;
- read examples just once and in their order of arrival;
- use of a fixed amount of memory (independently of the number of examples processed);
- able to predict at any time (can be interrupted);
- able to adapt the model in case of concept drift.

Similar properties were presented in [46] but the study was in the context of large data set and not of streaming data. More recently, [47] proposed eight properties to qualify an algorithm for streaming data. However this last article is more related to database field than to data mining. Table 1 summarizes the expected properties in these three papers.

Table 1. Expected properties for a learning algorithm on data streams: (1) = [46]; (2) = [48]; (3) = [47].

	(1)	(2)	(3)
Incremental	x	x	
Read data only once	x	x	x
Memory management	x	x	x
Anytime	x	x	x
Deal with concept drift		x	

Secondly, we focus on the criteria to compare the learning methods. Table 2 presents this comparison with the criteria generally used [49] to evaluate learning algorithm. The deployment complexity is given in this table for a two classes problem and represents the upper bound of the complexity to obtain one of the conditional probability for a test example: $P(C_k|X)$. This upper bound is rarely reached for many classifiers: for instance if we take a decision tree (and the tree is not fully unbalanced), the complexity is $O(h)$ where h represents the depth of the tree.

Finally the differences between incremental learning for batch learning or for data streams are presented in Table 3. A "post-optimization" step can be performed at the end of the learning process. This step aims to improve the model without necessarily reading again the examples.

Table 2. Comparison of the main algorithms: n number of examples; j number of attributes; a number of rules; b average number of predicates per rule, s number of support vectors.

Criterion	Decision tree	SVM	Nearest neighbour	Rules base system	Naive Bayes
Characteristics of the learning algorithm					
Learning speed	$+$	$--$	$++$	$-$	$+$
Deployment complexity	$O(a)$	$O(sj)$	$O(nj)$	$O(ab)$	$O(j)$
Speed and ability to update the model	$--$	$-$	$++$	$+$	$++$
CPU - memory	$+$	$+$	$--$	$+$	$+$
Relevance of the classifier					
Accuracy	$++$	$++$	$+$	$++$	$+$
Number of parameters	$-$	$-$	$++$	$+$	$++$
Speed to predict	$++$	$-$	$-$	$+$	$++$
Model visualization	$++$	$-$	$++$	$++$	$++$
Noise sensibility	$--$	$++$	$-$	$-$	$++$

Table 3. Properties of incremental learning vs incremental learning on streams (yes = required, no = not required).

	Incremental	Incremental on streams
Tuning of the learner settings using cross-validation	No	No
Data read just once	Yes	Yes
Post-optimization after learning	Yes	No
Complexity for learning and prediction	Low	Very low
Memory management	Yes	Yes
Handling of the trade-off between accuracy and time to learn	No	Yes
Concept drift handling	No	Yes
Anytime	No	Recommended

The next parts in this section present algorithms which are incremental and also seem to be adapted to process data streams: able to learn without slowing down the stream (lower complexity than the algorithms in the previous section), able to reach a good trade-off between processing/memory/precision, able to deal with concept drifts. They fulfill all or a part of the criteria presented in Table 3.

4.2   Decision Tree

Introduction

Limits of the existing algorithms: SLIQ [50], SPRINT [51], RAINFOREST [52] are algorithms specifically designed to deal with very large data sets. However they require reading many times the data and therefore they do not fulfill the stream constraint of reading only once the data. Few algorithms are able to see data just once as for instance ID5R [17] or ITI [18]. Nevertheless sometimes it is more expensive to update the model than to learn it from scratch, hence these algorithms cannot be considered as "online" or "anytime".

Tree size: Decision trees are built in a way that makes them always grow as new data arrive [53] but a deeper tree usually means over-fitting than better performance ([49] - p.203). In the context of stream mining, a tree would never stop to grow if there are no mechanisms to prune it. Therefore new algorithms were proposed to build trees on a data stream that will have dedicated mechanisms to stop them from growing.

Concept drift: Data streams usually have concept drift and in that case the decision tree must be pruned or restructured to adapt to the new concept [17,18].

Hoeffding bound: Numerous incremental learning algorithms on streams use the Hoeffding bound [54] to find the minimal number of examples needed to split a leaf into a node. For a random variable r within the range R and after n independent observations, the Hoeffding bound states that, with a probability $1 - \delta$, that the true mean of a r is not different from $\bar{r} - \epsilon$, where $\epsilon = \sqrt{\frac{R^2}{2n} ln(\frac{1}{\delta})}$. This bound does not depend on the data distribution but only on:

- the range R
- the number of observations n
- the desired confidence δ

As a consequence, this bound is more conservative since it has no prior on the data distribution[1].

Main Algorithms in the Literature: This section presents the main incremental decision trees for data streams in the literature:

VFDT [56] is considered as a reference article for learning on data streams with millions/billions of examples. This article is widely referenced and compared to new approaches on the same problems since 2000. In VFDT, the tree is built incrementally and no examples are kept. The error rate is higher in the early stage of the learning than an algorithm as C4.5. However after processing millions

---

[1] This bound is not well used in many algorithms of incremental trees as explain in [55] but with not a very big influence on the results.

of examples, the error rate become lower than C4.5 which is not able to deal with millions of examples but has to use just a subset of them. Figure 1, taken from VFDT article, shows this behavior and why it is better to use VFDT in the case of massive data streams. Moreover Domingos and Hulten proved that the "Hoeffding Tree" is similar to the tree that would have been learned with an off-line algorithm. In order to suit better the use case of stream mining VFDT can be tuned. The two main parameters are: (i) the maximum amount of memory to use, (ii) the minimal number of examples seen before calculating the criterion.

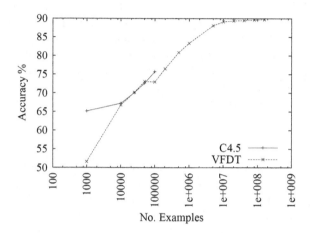

Fig. 1. Comparison of C4.5 with VFDT (taken from [56])

CVFDT [48] is an extension of VFDT to deal with concept drift. CFVDT grows an alternative subtree whenever an old one seems to be out-of-date, and replaces the old subtree when the new one becomes more accurate. This allows one to make adjustments when concept drift occurs. To limit the memory usage only the promising sub-trees are kept. On the dataset "rotating hyperplane" (see Sect. 6.3), the experiments show that the tree contains four times less nodes but the time spent to build it is five times longer comparatively to VFDT.

VFDTc [57] is an extension of VFDT able to deal with numerical attributes and not only categorical. In each leaf and for each attribute counters for numerical values are kept. These counters are stored using a binary tree so that it is efficient to find the best split value for each attribute to transform this leaf into a node. The authors of VFDTc observe that it needs 100 to 1,000 examples to transform a leaf into a node. In order to improve the performance of the tree, they propose to add a local model in the leaves. The naive Bayes classifier is known to have good performance with few data (Sect. 3.5) and therefore this classifier is used in the leaves of VFDTc. Local models improve the performance of the tree without using extra amount of memory since all the needed statistics are already available.

In **IADEM** [58] and **IADEMc** [59] the tree construction is also based on the Hoeffding bound. The tree growth is stopped using the error rate given as a parameter to the algorithm. IADEMc is an extension able to handle numerical attributes and which have a naive Bayes classifier in its leaves (as VFDTc). Their experiments on the WaveFrom and LED datasets show the following differences: a slightly lower accuracy but less deep trees.

Finally Kirkby in [60] makes a study on "Hoeffding Tree" and proposes improvements. His first improvement concerns leaves that could have either a naive Bayes or a majority class classifier. His experiments show that having a local classifier is not always the best option. He proposes to use the classifier which has the lowest error rate in its leaf. His second improvement is "Hoeffding Option Trees" (related to [61]) which have few sub-trees in their nodes. The same learning examples can update many trees at the same time. The prediction is a majority vote to predict the class value. Bagging [62] and boosting [63] techniques are also studied: bagging surely improves the accuracy but boosting does not always give a positive result.

Discussion

Limits of decision trees: Because of the tree structure its first node is very important. If a change occurs on the attribute on top nodes then it is more complicated to update the tree. In that case, it might be needed to rebuild the whole tree from scratch as stated in [64]. In that case, using decision trees for classification on non-stationary streams can be costly. The ability for an incremental tree to get restructured is for sure a key point to have good performance.

About the "anytime" aspect, [65] proposed to use a tree containing a Bayesian model in all the nodes. If there is not enough time to reach a leaf, the model in the last visited node is used.

About the Hoeffding bound: one could mention that other bounds like the McDiarmid bound exist and could be used as well. In [66] Rutkowski et al. show that the Hoeffding's inequality is not totally appropriate to solve the underlying problem. They prove two theorems presenting the McDiarmid's bound for both the information gain, used in ID3 algorithm, and for Gini index, used in Classification and Regression Trees (CART) algorithm.

4.3   Support Vector Machine (SVM)

In [67], a SVM version to learn on large datasets is presented. This version is based on an approximation of the optimal solution using "MEB - Minimum Enclosing Balls" resolution. Its spatial complexity is independent of the learning dataset size. A parameter tunes the trade-off between speed and precision of the classifier.

Reference [68] presents an optimization of the existing methods in order to increase their throughput. This optimization use the "divide and conquer" technique to parallelize the problem into sub problems. The goal is to rapidly find and exclude examples which are not support vectors.

We could also use LASVM [24–26] on each example of the stream: either the example is close to the actual support vector and the current solution is updated; or it is far and then the example is ignored. If the example is used, the complexity is in the best case $O(s^2)$ with s the number of support vectors. In both cases (example ignored or not), the elements from the kernel need to be computed. This pretty high complexity can explain the very few use cases of "LASVM" in data stream mining (compared to Hoeffding trees). Reference [69] presents guarantees for "LASVM" in the context of online learning. To improve the training time to build the SVM, [70] proposed to use linear SVM and parallelize computation using GPUs to reduce the computation of a factor 100.

4.4 Rule-Based System

Few articles proposed rule-based systems on data streams. Incremental rule-based algorithms exists (Sect. 3.4) but they are unable to deal with data streams. Nevertheless three articles of Ferrer et al. [71,72] proposed the FACIL algorithm able to deal with data streams. FACIL is based on the rule-based system AQ-PM [30].

The FACIL algorithm works in the following way:

- when a new example arrives, it looks for all the rules covering and having its label, then it increments the rule positive support.
- if the example is not covered then it looks for the rule which needs the minimal increase of its coverage space. Moreover this increase has to be limited to a given value κ to be accepted.
- if no rule with the same label as the example are found, then it looks into the whole set of rules with a different label. If a rule is found then it adds this example as a negative one and increments the rule negative support.
- if no rule covers this example, then a new rule can be created.
- each rule has its own forgetting window. This window has a variable size.

More recently Gama and Kosina [73] proposed a new algorithm based on the Hoeffding bound to build rules. A rule is expanded if, for an attribute, the gain to split on an attribute is better than not to split (the notion of gain is the same as in VFDT). Moreover another condition is added: the value of the chosen attribute has to be seen in more than 10 % of the examples. Similarly to VFDTc, they add a naive Bayes classifier in each rule to improve their accuracy.

4.5 Naive Bayes

The naive Bayes classifier uses conditional probabilities estimates (Sect. 3.5). These estimates are usually done after a discretization of the explicative variables. In the stream mining context this first step needs dedicated methods as most off-line methods usually need to load all the data in memory. In the literature, two kinds of publication related to incremental discretization can be found. The articles related to data-mining are not numerous but the literature related to database provides much more articles because the database systems (DBMS) need to have good estimates of the data distribution.

Discretization for Data Mining: Gama and Pinto in [74], propose PiD: "Partition Incremental Discretization" which performs an incremental discretization in order to build a naive Bayes classifier for streams. This discretization is based on two layers:

- first layer: it aims to lower the spatial and processor complexity of the streams. It stores per interval the number of examples per class. These intervals evolve with the stream using the following algorithms: if new values arrive and are outside the range of data already arrived then a new interval is inserted (∼EqualWidth); if an interval contains too many examples then this interval is split into two intervals (∼EqualFreq). The update is incrementally done. This first layer needs to contain much more intervals than the second layer. The number of intervals in the first layer is one of the parameters of the algorithm.
- second layer: it aims to have a more qualitative discretization to build a good model. It takes as an input the discretization of the first layer and builds the final discretization which can be: EqualWidth, EqualFreq, K-means, Proportional discretization, Recursive entropy discretization, MDL.

One issue that could arise with this method is due to the first layer that can use more memory than expected. This could happen if the data distribution is much skewed or in the presence of numerous outliers. In that case lots of new intervals will be created.

Discretization for DBMS: Database systems (DBMS) also used discretization for estimating data distribution. These estimations are used to have statistics on the data in order to build accurate execution plans and know which optimization/join to use. For DBMS it is necessary that these algorithms are able to deal with insertion and deletion but for data-mining on data streams this is not always mandatory.

Gibbons et al. [75] propose new incremental algorithms to update the histograms for database systems. Their approach is based on two structures:

- a reservoir to store representative examples (based on "Reservoir sampling" [76]);
- "Equal Frequency" histograms to summarize the data.

Weighted Naive Bayes: After the computation of the conditional densities $(log(P(X|C)))$ a weight on each explanatory variable could be computed online as in [77]. These kinds of method allow the naive bayes to reach better results.

4.6 KNN - Lazy Learning

Nearest neighbors algorithms are easily incremental (Sect. 3.6). In the context of stream meaning, most of approaches focus on keeping the most representative examples so that the memory consumption does not increase. This can be achieved by (i) forgetting the oldest examples, (ii) aggregating the nearest neighbors.

Different methods based on the nearest neighbors are proposed in the literature for stream mining:

- [78] used a discretization technique to lower the input space and therefore limit the memory needed to store the examples.
- [79] keeps the most recent examples and used a forgetting window to update them.
- [80] considers that the key problem for learning on data stream is to maintain an implicit concept description in the form of a case base (memory).

4.7   Other Methods

The main approaches in the literature for incremental classification presented above do not cover all the literature of online learning. A general scheme for others classifiers could be found in [81]. We give below a list of other online classifiers (which do not necessarily have all the properties given in Table 1):

Examples inspired on The Immortal Perceptron [82,83]: Second-Order Perceptron [84], Ultraconservative Algorithms [85], Passive-Aggressive Learning [86].

Examples based on Stochastic Optimization: Pegasos [87], Stochastic Gradient Descent [88].

Examples based on Recursive Least Squares/Optimal Control: Online Kernel Recursive Least Squares [89], Sparse Online Gaussian Processes [90].

Example based on Bandits[2]: Thompson Sampling [91], UCB, UCT and variants [92], Exp3, and variants [93].

Examples (Online Learning with Kernels) dedicated to deal with concept drift (see Section 5): Kernel Perceptron [94], Passive-Aggressive Learning [86] (implemented in Mahout), Pegasos [87], Budget online learning (Budget Perceptron [95], Forgetron [96], Projectron [97]).

---

[2] Multi-armed bandits explore and exploit online set of decisions, while minimizing the cumulated regret between the chosen decisions and the optimal decision. Originally, multi-armed bandits have been used in pharmacology to choose the best drug while minimizing the number of tests. Today, they tend to replace A/B testing for web site optimization (Google analytics), they are used for ad-serving optimization. They are well designed when the true class to predict is not known: for instance, in some domains the learning algorithm receives only partial feedback upon its prediction, i.e. a single bit of right-or-wrong, rather than the true label.

5   Concept Drift Management

Concept drift consists of changes in statistical properties of incoming examples. It causes a decrease in classifiers accuracy as time passes (non stationary data stream).

In supervised classification, the concept to learn $P(C|X)$ is the conditional probability of the class of C knowing the data observations X. A data stream could be non-stationary if the process which generates the data evolves over the time. In this case the classifier has to adapt gradually as the concept changes.

The literature on concept drift detection or concept drift management is abundant [98,99]. The methods on drift management can be split into several groups: drift detection, ensemble of classifiers, samples weighting, etc.

In this section we consider that the concept to learn may change over time but it is persistent and consistent (see [100]) between two changes. The period which exists between two changes is named context (see Sect. 2.1 [8]). Concept drift appears through the examples: the old observations are obsolete when considering the actual process which generates the actual observations. The old observations do not belong to the actual context.

We assume that between two context changes there exists a concept sufficiently persistent and consistent to be able to collect training examples. In that case concept drift management is mainly based on concept drift detection.

If a concept drift detection method is available, then we could:

– either retrain the decision model from scratch;
– or adapt the current decision model;
– or adapt a summary of the data stream on which the model is based;
– or work on a sequence of decision models which are learned over the time.

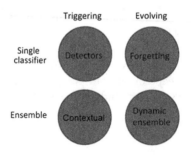

Fig. 2. Adaptive learning strategies [101].

In [101] another taxonomy is given by considering two types of method: (i) Triggering/Evolving and (ii) Single classifier/Ensemble of classifiers (see Fig. 2). In this figure, in case of:

– Single classifier + Triggering: detect a change and cut;
– Single classifier + Evolving: forget old data and retrain;

- Ensemble of classifiers + Triggering: build many models, switch models according to the observed incoming data;
- Ensemble of classifiers + Evolving: build many models, dynamically combine them.

The literature indicates that concept drift may have different faces. Assuming that the concept of a sequence of instances changes from S_i to S_j, we call a drift a sudden drift, if at time t, S_i is suddenly replaced by S_j. Instead of a sudden change which means that the source is switched during a very short period of time after the change point t, gradual drift refers to a change with a period both sources S_i and S_j are active. As time passes, the probability of sampling from source S_i decreases, probability of sampling from source S_j increases until S_i is completely replaced by S_j. There is another type of gradual drift called incremental or stepwise drift. With more than two sources, the difference between the sources is very small, thus the drift is noticed only when looking at a longer time period. Reoccurring context means that previously active concept reappears after some time. It is not certainly periodic. So there is not any predictive information on when the source will reappear. In Fig. 3, we can see how these changes happen in a data stream.

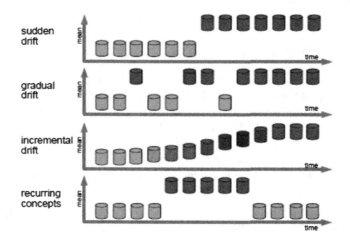

Fig. 3. Types of concept drift [99]

The goal of the supervised classification on data streams is to keep the performance of the classifier over the time, while the concept to learn is changing. In this case, the literature contains three families:

- methods without a classifier: authors mainly use the distributions of the explanatory variables of the data stream: $P(X)$, $P(C)$, $P(X|C)$.
- methods using a classifier: authors mainly use the performance of the classifier related to the estimation of $P(C|X)$.
- adaptive methods with or without using a classifier.

From a Bayesian viewpoint, all these methods try to detect the variation of a part of:

$$P(C|X) = P(C)P(X|C)/P(X) \tag{1}$$

with

1. $P(C)$ the priors of the classes in the data stream
2. $P(X)$ the probability (distribution) of the data
3. $P(X|C)$ the conditional distribution of the data X knowing the class C.

The following section is organized in two parts: the first part describes methods to perform concept drift detection; the second part describes methods to adapt the model (single classifier or ensemble of classifiers). In this section we do not consider the novelty detection problem [102]. Novelty detection is a useful ability for learning systems, especially in data stream scenarios, where new concepts can appear, known concepts can disappear and concepts can evolve over time. There are several studies in the literature investigating the use of machine learning classification techniques for novelty detection in data streams. However, there is no consensus regarding how to evaluate the performance of these techniques, particular for multiclass problems. The reader may be interested by a new evaluation approach for multiclass data streams novelty detection problems described in [103].

5.1   Concept Drift Detection

An extensive literature on concept drift exists. The interested reader may find in [99] an overview of the existing methods. In the following subsections we present only a part of the literature related to drift detection methods applied on one of these three terms: $P(C)$, $P(X)$ and $P(X|C)$.

Concept drift detection is mainly realized by (i) monitoring the distributions of the explanatory variables of the data stream: $P(X)$, $P(C)$, $P(X|C)$; (ii) monitoring the performance of the classifier related to the estimation of $P(C|X)$.

Methods Using a Classifier: Methods using a classifier monitor the performance of the classifier and detect a concept drift when the performance varies significantly. These methods assume that the classifier is a stationary process and data are independent and identically distributed (iid). Though, these hypothesis are not valid in case of a data stream [55], these methods proved their interest on diverse application [98, 104, 105]. The three most popular methods using a classifier of the literature are described below.

– **Widmer and Kubat** [29] proposed to detect the change in the concept by analyzing the misclassification rates and the changes that occur in the structures of the learning algorithm (adding new definitions in the rules in his case). From some variation of these indicators, the size of the training window is reduced otherwise this window grows in order to have more examples to achieve better learning. This article was one of the first works on this subject. Many other authors, like Gama in the paragraph below, based their methods on it.

- **DDM** This method of Gama et al. [104], detects a concept drift by monitoring the classifier accuracy. Their algorithm assumes that the binary variable which indicates that the classifier has correctly classified the last example follows a binomial distribution law. This law can be approximated as a normal law when the number of examples is higher than 30. The method estimates after the observation of each sample of the data stream the probability of misclassification p_i (p_i corresponds also to the error rate) and the corresponding standard deviation $s_i = \sqrt{p_i(1 - p_i)/i}$. A significant increase of the error rate is considered by the method as the presence of a concept drift.

  The method uses two decision levels: a "warning" level when $p_i + s_i \geqslant p_{min} + 2 \cdot s_{min}$ and a "detection level" when $p_i + s_i \geqslant p_{min} + 3 \cdot s_{min}$ (after the last detection, every time a new example i is processed p_{min} and s_{min} are updated simultaneously according to $(p_{min} + s_{min}) = \min_i(p_i + s_i)$). The time spent between the warning level and the detection level is used to train a new classifier which will replace the current classifier if the detection level is reached. This mechanism allows not starting from scratch when the decision level is reached (if the concept drift is not too sudden).

- **EDDM** This method [105] uses the same algorithm but with another criterion to set the warning and detection levels. This method uses the distance between the classification errors rather the error rate. This distance corresponds to the number of right predictions between two wrong predictions. EDDM computes the mean distance between the errors p_i' and the corresponding standard deviation s_i' As for DDM a warning level and a detection level are defined, respectively of $(p_i' + 2 \cdot s_i')/(p_{max}' + 2 \cdot s_{max}') < \alpha$ and $(p_i' + 2 \cdot s_i')/(p_{max}' + 2 \cdot s_{max}') < \beta$. In the experimental part the authors of EDDM use $\alpha = 90\%$ and $\beta = 95\%$. On synthetic datasets EDDM detects faster than DDM the gradual concept drift.

Methods Without a Classifier: The methods without a classifier are mainly methods based on statistical tests applied to detect a change between two observation windows. These methods are often parametric and need to know the kind of change to detect: change in the variance, change in the quantiles, change in the mean, etc. A parametric method is a priori the best when the change type exactly corresponds to its setting but might not perform well when the problem does not fit its parameters. In general, it is a compromise between the particular bias and the effectiveness of the method. If one seeks to detect any type of change it seems difficult to find a parametric method. The list below is not exhaustive but gives different ideas on how to perform the detection.

- **Welch's t test:** this test applies on two samples of data of size N_1 and N_2 and is an adaptation of the Student's t test. This test is used to statistically test the null hypothesis that the means of two population $\overline{X}_1$ and $\overline{X}_2$, with unequal variances (s_1^2 and s_2^2), are equals. The formula of this test is:

$$p\text{-}value = (\overline{X}_1 - \overline{X}_2) / \left(\sqrt{(s_1^2/N_1) + (s_2^2/N_2)} \right) \tag{2}$$

The null hypothesis can be rejected depending on the $p\text{-}value$.

- **Kolmogorov-Smirnov's test:** this test is often used to determine if a sample follow (or not) a given law or if two samples follow the same law. This test is based on the properties of their empirical cumulative distribution function. We will use this test to check if two samples follow the same law. Given two samples of size N_1 and N_2 having respectively cumulative distribution function $F_1(x)$ and $F_2(x)$, the Kolmogorov-Smirnov distance is defined as: $D = \max_x |F_1(x) - F_2(x)|$. The null hypothesis, assuming that the two samples follow the same law, is rejected with a confidence of α if: $\left(\sqrt{(N_1N_2)/(N_1 + N_2)} \right) D > K_\alpha$. K_α can be found in the Kolmogorov-Smirnov table.

- **Page-Hinkley test:** Gama in [106] suggests to use the Page-Hinkley test [107]. Gama modifies this test using fading factor. The Page-Hinkley test (PHT) is a sequential analysis technique typically used for monitoring change detection. It allows efficient detection of changes in the normal behavior of a process which is established by a model. The PHT was designed to detect a change in the average of a Gaussian signal [108]. This test considers a cumulative variable defined as the cumulated difference between the observed values and their mean till the current moment.

- **MODL** $P(W|X_i)$ this method has been proposed by [109] to detect the change while observing a numerical variable X_i in an unsupervised setting. This method addresses this problem as a binary classification problem. The method uses two windows to detect the change: a reference window W_{ref} which contains the distribution of X_i at the beginning and a current (sliding or jumping) window W_{cur} which contains the current distribution. The examples belonging to W_{ref} are labeled with the class "ref" and the ones belonging to W_{cur} are labeled with the class "cur". This two labels constitute the target variable $W \in \{W_{ref}, W_{cur}\}$ of the classification problem. All these examples are merged into a training set and the supervised MODL discretization [110] is applied to see if the variable X_i could be split in more than a single interval. If the discretization gives at least two intervals then there are at least two significantly different distributions for X_i conditionally to the window W. In this case the method detects that there is a change between W_{ref} and W_{cur}.

5.2   Adaptive Methods

The online methods can be divided into two categories [111] depending on whether or not they use an explicit concept drift detector in the process of adaptation to evolving environments.

In case of explicit detection, after a concept drift was detected we have to: (i) either retrain the model from scratch; (ii) or adapt the current decision model;

(iii) or adapt a summary of the data stream on which the model is based; (iv) or work on a sequence of models which are learned over the time.

If we decide to retrain the classifier from scratch the learning algorithm uses:

- either a partial memory of the examples: (i) a finite number of examples; (ii) an adaptive number of examples; (iii) a summary of the examples. The size of the window of the last used examples cannot exceed the last detected concept drift [112].
- or statistics on all the examples under the constraint of memory and/or the precision on the statistics computed [113]. In this case, the idea is to keep only in memory the "statistics" is useful to train the classifier.

In case of implicit adaptation, the learning system adapts implicitly to a concept change by holding an ensemble of decision models. Two cases exist:

- either the predictions are based on all the classifiers of the ensemble using simple voting, weighted voting, etc. [114]:

  • SEA (Streaming Ensemble Algorithm) [115]: the data stream fills a buffer of a predefined size. When the buffer is full a decision tree is built by the C4.5 algorithm on all the examples in the buffer. Then new buffers are created to repeat the process and these classifiers are put in a pool. When the pool is full: if the new inserted classifier improves the prediction of the pool then it is kept, otherwise it is rejected. In [64] the same technique is used, but different types of classifiers are used in the pool to incorporate more diversity: Bayesian naive, C4.5, RIPPER, etc.
  • In [116], a weight is given to each classifier. The weights are updated only when a wrong prediction of the ensemble occurs. In this case a new classifier is added to the ensemble and the weights of the others are decreased. The classifiers with too small weights are pruned.
  • ADWIN (ADaptative WINdowing) proposed by [98,117] detects a concept change using an adaptive sliding window model. This method uses a reservoir (a sliding window) of size W which grows up when there is no concept drift and decrease when there is a concept drift. The memory used is $log(W)$. ADWIN looks at all possible sub-windows partitions in the window of training data. Whenever two large enough sub-windows have distinct enough averages, a concept change is detected and the oldest partition of the window is dropped. Then the worst classifier is dropped and a new classifier is added to the ensemble. ADWINs only parameter is a confidence bound, indicating how confident we want to be on the detection.
  • ADACC (Anticipative Dynamic Adaptation to Concept Changes) [118] is a method suggested to deal with both the challenge of optimizing the stability/ plasticity dilemma and with the anticipation and recognition of incoming concepts. This is accomplished through an ensemble method that controls an ensemble of incremental classifiers. The management of the ensemble of classifiers naturally adapts to the dynamics of the concept changes with very few parameters to set, while a learning mechanism managing the changes in the ensemble provides means for anticipation and quick adaptation to the underlying modification of the concept.

– or the predictions is based on a single classifier of the ensemble: they are used one by one and other classifiers are considered as a pool of potential candidates [119, 120].

Another way is to weight along the time the examples used to train the decision model. Since the 80's several methods have been proposed as the fading factors [28, 29, 112, 121]. This subject remains relevant for online algorithms. We can cite for example [121]:

– TWF (Time-Weighted Forgetting): the older the example is, the lower is its weight;
– LWF (Locally-Weighted Forgetting): at the arrival of a new example the weights of its closest examples are increased and the weights of the other examples are decreased. The region having recent examples are then kept or created if they did not exist. Regions with no or few examples are deleted. This kind of method is used also for unsupervised learning as in [122];
– PECS (Prediction Error Context Switching): the idea is the same as LWF but all examples are stored in memory and a checkup of the labels of the new examples over the old ones is done. A probability is calculated using the number of examples having or not the same labels. Only the best examples are used to achieve learning.

A recent survey on adaptive methods may be found in [118] and the reader may also be interested in the book [123], the tutorial given during the PAKDD conference [101] or the tutorial given at ECML 2012 [124].

6   Evaluation

In the last years, many algorithms have been published on the topic of supervised classification for data streams. Although Gama in [106] and then in [8] (Chap. 3) suggests an experimental method to evaluate and to compare algorithms for data streams. The comparison between algorithms is not easy since authors do not always use the same evaluation method and/or the same datasets. Therefore the aim of this section is to give, in a first part, references about the criteria used in the literature and in a second part references on the datasets used to perform comparisons. Finally, in a third part we discuss other possible points of comparison.

6.1   Evaluation Criteria

The evaluation criteria are the same as those used in the context of off-line learning, we can cite the main ones:

– Accuracy (ACC): the accuracy is the proportion of true results (both true positives and true negatives) in the population. An accuracy of 100 % means that the measured values are exactly the same as the given values.
– Balanced Accuracy (BER): the balanced accuracy avoids inflated performance estimates on imbalanced datasets. It is defined as the arithmetic mean of sensitivity and specificity, or the average accuracy obtained on either class.

- Area Under the ROC curve (AUC): Accuracy is measured by the area under the ROC curve [125]. An area of 1 represents a perfect test; an area of 0.5 results from a classifier which provides random prediction.
- Kappa Statistic (K) where $K = \frac{(p_0 - p_c)}{(1 - p_c)}$ with p_0: the prequential accuracy of the classifier and p_c: the probability that a random classifier makes a correct prediction. K = 1 if the classifier is always correct and K = 0 if the predictions coincide with the correct ones as often as those of the random classifier.
- Kappa Plus Statistic (K^+): a modification a of the Kappa Statistic has recently been presented in [126,127] where the random classifier is replaced by the "persistent classifier" (which predict always the label of the latest example received).

6.2   Evaluation Methods

In the context of off-line learning the most used evaluation method is the cross validation named k-fold cross validation. This technique is a model validation technique for assessing how the results of a statistical analysis will generalize to an independent data set. The goal of a cross validation is to define a dataset to "test" the model in the training phase (i.e., the validation dataset), in order to limit over-fitting, give an insight on how the model will generalize to an independent data set (i.e., an unknown dataset, for instance from a real problem), etc.

One round of cross-validation involves partitioning a sample of data into disjoint subsets, learning the classifier on one subset (called the training set), and validating the classifier on the other subset (called the validation set or testing set). To reduce variability, multiple rounds of cross-validation (k) are performed using different partitions, and the validation results are averaged over the rounds.

But the context is different in the case of online learning on data streams. Depending on whether the stream is stationary or not (presence of concept drift or not) two main techniques exist:

- "Holdout Evaluation" requires the use of two sets of data: the train dataset and the holdout dataset. The train dataset is used to find the right parameters of the classifier trained online. The holdout dataset is then used to evaluate the current decision model, at regular time intervals (or set of examples). The loss estimated in the holdout is an unbiased estimator. The constitution of the independent test set is realized a the beginning of the data stream or regularly (see Fig. 4).
- "Prequential Evaluation" [106]: the error of the current model is computed from the sequence of examples [128]. For each example in the stream, the actual model makes a prediction based only on the example attribute-values. The prequential-error is the sum of a loss function between the prediction and observed values (see Fig. 5). This method is also called: "Interleave Test-Then-Train". One advantage of this method is that the model is always being tested on unseen examples and no holdout set is needed so that available data are fully used. It also produces smooth plot accuracy over time. One may find prequential evaluation of the AUC in [129].

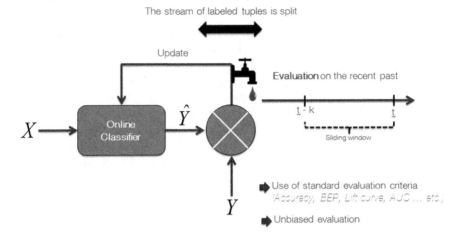

Fig. 4. Holdout Evaluation

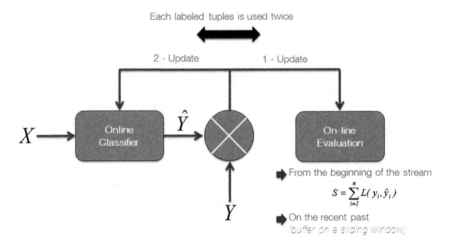

Fig. 5. Prequential Evaluation

Figure 6 compares the errors rates $(1 - Accuracy)$ for the algorithm VFDT on the WaveForm dataset (for which the error rate is bounded by an optimal Bayes error of 0.14) using (i) the holdout evaluation and the prequential evaluation [106].

The use of fading factors allows a less pessimistic evaluation as Gama suggested in [106] and as illustrated in Fig. 7. The use of a fading factor decreases the weights of predictions from the past.

In case of:

- stationary data streams: the Holdout or Prequential evaluation can be used, though the Holdout Evaluation is the most used in the literature;
- presence of concept drifts: the Holdout Evaluation cannot be used since the test and train dataset could incorporate examples from distinct concepts.

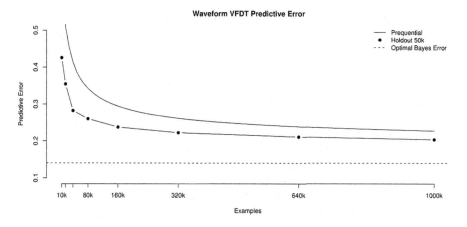

Fig. 6. Comparison between the Holdout Evaluation and the Prequential Evaluation

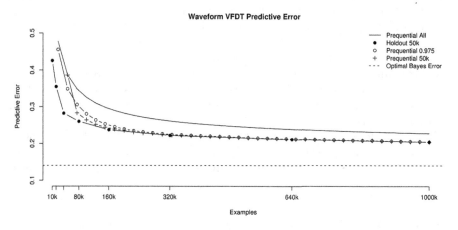

Fig. 7. Comparison between the Holdout Evaluation and the Prequential Evaluation using fading factor

Unlike the Holdout Evaluation, the Prequential Evaluation can be used and is adapted since its test dataset will evolve with the data stream.

6.3 Datasets

In many articles, the UCI Machine Learning Repository is cited as a source of real world datasets. This allows having a first idea of the performance of a new algorithm. To simulate a data stream several approaches can be considered depending on whether the data stream should contain a drift or not. We present below several data streams used in scientific publications. The reader can find other data streams in [8] p. 209.

Synthetic Data Generators without Drift: The best way to have enough data to test online algorithm on data streams, synthetic data generators exist to generate billions of examples. A list and description of these generators is presented in [60]. The main generators are:

- Random RBF [130] creates numerical dataset whose classes are represented in hyper sphere of examples with random centers. Each center has a random position, a single, standard deviation, class label and weight. To create a new example, a direction and a distance are chosen to move the example from the central point. The distance is randomly drawn from a Gaussian distribution with the given centroid standard deviation. The population in a hyper sphere depends on the weight of the centroid.
- Random Tree Generator: produces a decision tree using user parameters and then generates examples according to the decision tree. This generator is biased in favor of decision trees.
- LED Generator [14]: (LED Display Domain Data Set) is composed of 7 Boolean attributes which are predicted LED displays (the light is on or not) and 10 concepts. Each attribute value has the 10 % of its value inverted. It has an optimal Bayes classification rate of 74 %.
- Waveform (Waveform Database Generator Dataset) [14] produces 40 attributes and contains noise. The last 19 attributes are all noise attributes with mean 0 and variance 1. It differentiates between 3 different classes of waves, each of which is generated from a combination of two or three base waves. It has an optimal Bayes classification rate of 86 %.
- Function generator [131] produces a stream containing 9 attributes (6 numeric and 3 categorical) describing hypothetical loan applications. The classes (whether the loan should be approved) are presented in a Boolean label defined by 10 functions.

Synthetic Data Generators with Drift: to test the performance of their online algorithms when there is a concept drift, authors suggested synthetic data generators where the user may include a drift where the position, the size, the level, etc. can be parameterized (see also [8] p. 209). The previous generators may also be changed to contain changes in their data stream.

- SEA Concept generator [115] contains abrupt drift and generates examples with 3 attributes between 0 and 10 where only the first 2 attributes are relevant. These examples are divided into 4 blocks with different concepts by giving each block a threshold value which is bound by the sum of the first two attributes. This data stream has been used in [132].
- STAGGER [28] is a collection of elements, where each individual element is a Boolean function of attribute-valued pairs represented by a disjunct of conjuncts. This data stream has been used in [79, 98, 116, 133].
- Rotating hyperplane [48] uses a hyperplane in d-dimension as the one in SVM and determines the sign of labels. It is useful for simulating time-changing concepts, because we can change the orientation and position of the hyperplane in a smooth manner by changing the relative size of weights. This data stream has been used in [64, 71, 79, 98].

– Minku's artificial problems: a total of 54 datasets were generated by Minku et al. in 2010, simulating different types of concept changes [134]. A concept change event is simulated by changing one or more of the problems parameters, creating a change in the conditional probability function.

Real Data: in these datasets a change in the distribution (concept drift or covariate shift or both) could exist but the position is not known. Because of their relatively large size they are used to simulate a real data stream:

– Proxy web of the University of Washington [48, 56].
– Electricity dataset (Harries, 1999) [132]: This data was collected from the Australian New South Wales Electricity Market. It contains 45,312 examples drawn from 7 May 1996 to 5 December 1998 with one example for each half hour. The class label (DOWN or UP) identifies the change of the price related to a moving average of the last 24 h. The attributes are time, electricity demands and scheduled power transfer between states, all of which are numeric values.
– Forest Cover type dataset (Covertype) contains 581012 instances of 54 categorical and integer attributes describing wildness areas condition collected from US Forest Service (USFS) Region 2 Resource Information System (RIS) and US Geological Survey (USGS) and USFS data. The class set consists of 7 forest cover types.
– Poker-Hand dataset (Poker Hand): each record is an example of a hand consisting of five playing cards drawn from a standard deck of 52. Each card is described using two attributes (suit and rank), for a total of 10 predictive attributes. There is one Class attribute that describes the "Poker Hand". The order of cards is important, which is why there are 480 possible Royal Flush hands as compared to 4. This database has been used in [98].
– The KDD 99 cup dataset is a network intrusion detection dataset [135]. The task consists of learning a network intrusion detector able to distinguish between bad connections or attacks and good or normal connections. The training data consists of 494,021 examples. Each example represents a connection described by a vector of 41 features which contain both categorical (ex: the type of protocol, the network service) and continuous values (ex: the length of the connection, its duration). The class label is either 0 or 1 for normal and bad connection, respectively.
– The Airlines task[3] is to predict whether a given flight will be delayed, given the information of the scheduled departure. The dataset contains 539,383 examples. Each example is represented by 7 feature values describing the flight (airline, flight number, source, destination, day of week, time of departure and length of flight) and a target value which is either 1 or 0, depending on whether the flight is delayed or not. This dataset is an example of time-correlated stream of data. It is expected that when a flight is delayed (due to weather condition for instance), other timely close flights will be delayed as well.

---

[3] http://moa.cms.waikato.ac.nz/datasets/.

– The SPAM dataset [136] consists of 9,324 examples and was built from the email messages of the Spam Assassin Collection using the Boolean bag-of-words representation with 500 attributes. As mentioned in [51], the characteristics of SPAM messages in this dataset gradually change as time passes (gradual concept drift). This database is small but the presence of a known concept drift is interesting.

6.4 Comparison Between Online Classifiers

The same methodology is mainly used in the literature. At first, the authors of a new algorithm perform a comparison with known algorithms but not designed to operate on data streams as: C4.5, ID3, Naive Bayes, Random Forest, etc. The idea is to see the performance on small amounts of data against known algorithms. Then in a second time they confront against other algorithms already present in the literature on data streams. The reader can find in [60] and in [106] a comparison of evaluation methods between some of the algorithms discussed in this paper.

Platform for evaluation: one of the problems when we want to make a comparison between different algorithms is to easily perform experiments. Indeed, it is often difficult to obtain the source code of the algorithm then to successfully build an executable. Furthermore the input formats and output formats are sometimes different between the various algorithms. For off-line learning the Weka toolkit [137] proposed by the University of Waikato allows to quickly perform experiments. The same University proposed a toolkit to perform experiment on data streams for online learning: MOA [133]. MOA contains most of the data generators previously described and many online classifiers as the Hoeffding Trees.

Other points of comparison: the criteria presented in Sect. 6.1 are the mainly used in the context of supervised classification and off-line learning. But in the case of data streams other evaluation criteria could be used as:

– the size of the model: number of nodes for a decision tree, number of rules for rule-based system, etc.
– the learning speed: number of examples that could be learned per second,
– the return on investment (ROI): the idea is to asses and interprets the gains in performance due to model adaptation for a given learning algorithm on a given prediction task [138],
– the RAM-Hours used which is an evaluation measure of the resources used [139].

Comparing the learning speed of two algorithms can be problematic because it assumes that the code/software are publicly available. In addition, this comparison do not evaluates only the learning speed but also the platform on which runs the algorithm and the quality of its implementation. These measures are useful to give an idea but must be taken with caution and not as absolute elements of comparison.

Kirkby [60] realizes his experiments under different memory constraints simulating various environments: sensor network, PDA, server. The idea is therefore to compare different algorithms with respect to the environment in which they are executed.

We can also mention the fact that the precision measurement has no meaning in absolute terms in case of a drift in the data stream. The reader can then turn to learning protocols such as mistake-bound [140]. In this context, learning takes place in cycles where examples are seen one by one; which is consistent with learning on data streams. At the beginning the learning algorithm $(\mathcal{A})$ learns a hypothesis (f_t) and the prediction for the current instance is $f_t(x_t)$. Then the real label of the example (y_t) is revealed to the learning algorithm which may have made a mistake $(y_t \neq f_t(x_t))$. Then the cycle is repeated until a given horizon time (T). The error bound is related to the error maximum done on the time period T.

Limits of the evaluation techniques: new evaluation techniques have recently been proposed by the community and a consensus is looked for [106, 141]. Indeed, criteria such as the number of examples learned per second or the speed prediction depends on the machine, the memory used and the implementation quality. We must find the criteria and/or common platforms in order to achieve a comparison as impartial as possible.

7   Conclusion

This synthetic introductory article presented the main approaches in the literature for incremental classification. At first (Sect. 2.1) the various problems of learning were presented with the new challenges due to the constraints of data streams: quantity, speed and availability of the data. These characteristics require specific algorithms: incremental (Sect. 3) or specialized for data streams (Sect. 4). The main algorithms have been described in relation to their families: decision tree, naive Bayes, SVM, etc. The Sect. 5 focuses on processing non-stationary data streams that have many concepts. Learning on these streams can be handled by detecting changes in the concept or by having adaptive models. Finally, the last section addresses the evaluation in the context of data streams. New data generators based on drifting concept are needed. At the same time the standard evaluation metrics need to be adapted to the properly evaluate models built on drifting concepts.

Adapting models to changing concept is one of the most important and difficult challenge. Just few methods manage to find a compromise between accuracy, speed and memory. Writing this compromise as a criterion to optimize during the training of the classifier could be a path to explore. We can also note that data streams can be considered as a changing environment over which a classification model is applied. Misclassifications are then available in a limited time horizon. Therefore we can assume that the algorithms related to game theory or reinforcement learning could also be formalized for incremental learning on data streams.

This survey is obviously not exhaustive of all existing methods but we hope that it gives enough information and links on supervised classification for data streams to start on this subject.

References

1. Guyon, I., Lemaire, V., Dror, G., Vogel, D.: Analysis of the kdd cup 2009: fast scoring on a large orange customer database. In: JMLR: Workshop and Conference Proceedings, vol. 7, pp. 1–22 (2009)
2. Féraud, R., Boullé, M., Clérot, F., Fessant, F., Lemaire, V.: The orange customer analysis platform. In: Perner, P. (ed.) ICDM 2010. LNCS, vol. 6171, pp. 584–594. Springer, Heidelberg (2010)
3. Almaksour, A., Mouchère, H., Anquetil, E.: Apprentissage incrémental et synthèse de données pour la reconnaissance de caractères manuscrits en-ligne. In: Dixième Colloque International Francophone sur l'écrit et le Document (2009)
4. Saunier, N., Midenet, S., Grumbach, A.: Apprentissage incrémental par sélection de données dans un flux pour une application de securité routière. In: Conférence d'Apprentissage (CAP), pp. 239–251 (2004)
5. Provost, F., Kolluri, V.: A survey of methods for scaling up inductive algorithms. Data Min. Knowl. Discov. 3(2), 131–169 (1999)
6. Dean, T., Boddy, M.: An analysis of time-dependent planning. In: Proceedings of the Seventh National Conference on Artificial Intelligence, pp. 49–54 (1988)
7. Michalski, R.S., Mozetic, I., Hong, J., Lavrac, N.: The multi-purpose incremental learning system AQ15 and its testing application to three medical domains. In: Proceedings of the Fifth National Conference on Artificial Intelligence, pp. 1041–1045 (1986)
8. Gama, J.: Knowledge Discovery from Data Streams. Chapman and Hall/CRC Press, Atlanta (2010)
9. Joaquin Quinonero-Candela, J., Sugiyama, M., Schwaighofer, A., Lawrence, N.D.: Dataset Shift in Machine Learning. MIT Press, Cambridge (2009)
10. Bondu, A., Lemaire, V.: Etat de l'art sur les methodes statistiques d'apprentissage actif. RNTI A2 Apprentissage artificiel et fouille de données, 189 (2008)
11. Cornuéjols, A.: On-line learning: where are we so far? In: May, M., Saitta, L. (eds.) Ubiquitous Knowledge Discovery. LNCS, vol. 6202, pp. 129–147. Springer, Heidelberg (2010)
12. Zilberstein, S., Russell, S.: Optimal composition of real-time systems. Artif. Intell. 82(1), 181–213 (1996)
13. Quinlan, J.R.: Learning efficient classification procedures and their application to chess end games. In: Michalski, R.S., Carbonell, J.G., Mitchell, T.M. (eds.) Machine Learning - An Artificial Intelligence Approach, pp. 463–482. Springer, Heidelberg (1986)
14. Breiman, L., Friedman, J., Olshen, R., Stone, C.: Classification and Regression Trees. Chapman and Hall/CRC, Boca Raton (1984)
15. Cornuéjols, A., Miclet, L.: Apprentissage artificiel - Concepts et algorithmes. Eyrolles (2010)
16. Schlimmer, J., Fisher, D.: A case study of incremental concept induction. In: Proceedings of the Fifth National Conference on Artificial Intelligence, pp. 496–501 (1986)
17. Utgoff, P.: Incremental induction of decision trees. Mach. Learn. 4(2), 161–186 (1989)
18. Utgoff, P., Berkman, N., Clouse, J.: Decision tree induction based on efficient tree restructuring. Mach. Learn. 29(1), 5–44 (1997)
19. Boser, B., Guyon, I., Vapnik, V.: A training algorithm for optimal margin classifiers. In: Proceedings of the Fifth Annual Workshop on Computational Learning Theory, pp. 144–152. ACM, New York (1992)

20. Cortes, C., Vapnik, V.: Support-vector networks. Mach. Learn. **20**(3), 273–297 (1995)
21. Domeniconi, C., Gunopulos, D.: Incremental support vector machine construction. In: ICDM, pp. 589–592 (2001)
22. Syed, N., Liu, H., Sung, K.: Handling concept drifts in incremental learning with support vector machines. In: Proceedings of the Fifth ACM SIGKDD International Conference on Knowledge Discovery and Data Mining, pp. 317–321. ACM, New York (1999)
23. Fung, G., Mangasarian, O.: Incremental support vector machine classification. In: Proceedings of the Second SIAM International Conference on Data Mining, Arlington, Virginia, pp. 247–260 (2002)
24. Bordes, A., Bottou, L.: The Huller: a simple and efficient online SVM. In: Gama, J., Camacho, R., Brazdil, P.B., Jorge, A.M., Torgo, L. (eds.) ECML 2005. LNCS (LNAI), vol. 3720, pp. 505–512. Springer, Heidelberg (2005)
25. Bordes, A., Ertekin, S., Weston, J., Bottou, L.: Fast kernel classifiers with online and active learning. J. Mach. Learn. Res. **6**, 1579–1619 (2005)
26. Loosli, G., Canu, S., Bottou, L.: SVM et apprentissage des très grandes bases de données. In: Cap Conférence d'apprentissage (2006)
27. Lallich, S., Teytaud, O., Prudhomme, E.: Association rule interestingness: measure and statistical validation. In: Guillet, F., Hamilton, H. (eds.) Quality Measures in Data Mining. SCI, vol. 43, pp. 251–275. Springer, Heidelberg (2007)
28. Schlimmer, J., Granger, R.: Incremental learning from noisy data. Mach. Learn. **1**(3), 317–354 (1986)
29. Widmer, G., Kubat, M.: Learning in the presence of concept drift and hidden contexts. Mach. Learn. **23**(1), 69–101 (1996)
30. Maloof, M., Michalski, R.: Selecting examples for partial memory learning. Mach. Learn. **41**(1), 27–52 (2000)
31. Langley, P., Iba, W., Thompson, K.: An analysis of Bayesian classifiers. In: International Conference on Artificial Intelligence, pp. 223–228. AAAI (1992)
32. Domingos, P., Pazzani, M.: On the optimality of the simple Bayesian classifier under zero-one loss. Mach. Learn. **130**, 103–130 (1997)
33. Kohavi, R.: Scaling up the accuracy of naive-Bayes classifiers: a decision-tree hybrid. In: Proceedings of the Second International Conference on Knowledge Discovery and Data Mining, vol. 7. AAAI Press, Menlo Park (1996)
34. Heinz, C.: Density estimation over data streams (2007)
35. John, G., Langley, P.: Estimating continuous distributions in Bayesian classifiers. In. Proceedings of the Eleventh Conference on Uncertainty in Artificial Intelligence, pp. 338–345. Morgan Kaufmann (1995)
36. Lu, J., Yang, Y., Webb, G.I.: Incremental discretization for Naïve-Bayes classifier. In: Li, X., Zaïane, O.R., Li, Z. (eds.) ADMA 2006. LNCS (LNAI), vol. 4093, pp. 223–238. Springer, Heidelberg (2006)
37. Aha, D.W. (ed.): Lazy Learning. Springer, New York (1997)
38. Brighton, H., Mellish, C.: Advances in instance selection for instance-based learning algorithms. Data Min. Knowl. Discov. **6**(2), 153–172 (2002)
39. Hooman, V., Li, C.S., Castelli, V.: Fast search and learning for fast similarity search. In: Storage and Retrieval for Media Databases, vol. 3972, pp. 32–42 (2000)
40. Moreno-Seco, F., Micó, L., Oncina, J.: Extending LAESA fast nearest neighbour algorithm to find the k nearest neighbours. In: Caelli, T.M., Amin, A., Duin, R.P.W., Kamel, M.S., de Ridder, D. (eds.) SPR 2002 and SSPR 2002. LNCS, vol. 2396, pp. 718–724. Springer, Heidelberg (2002)

41. Kononenko, I., Robnik, M.: Theoretical and empirical analysis of relieff and rre-lieff. Mach. Learn. J. **53**, 23–69 (2003)
42. Globersonn, A., Roweis, S.: Metric learning by collapsing classes. In: Neural Information Processing Systems (NIPS) (2005)
43. Weinberger, K., Saul, L.: Distance metric learning for large margin nearest neighbor classification. J. Mach. Learn. Res. (JMLR) **10**, 207–244 (2009)
44. Sankaranarayanan, J., Samet, H., Varshney, A.: A fast all nearest neighbor algorithm for applications involving large point-clouds. Comput. Graph. **31**, 157–174 (2007)
45. Domingos, P., Hulten, G.: Catching up with the data: research issues in mining data streams. In: Workshop on Research Issues in Data Mining and Knowledge Discovery (2001)
46. Fayyad, U.M., Piatetsky-Shapiro, G., Smyth, P., Uthurusamy, R.: Advances in Knowledge Discovery and Data Mining. American Association for Artificial Intelligence, Menlo Park (1996)
47. Stonebraker, M., Çetintemel, U., Zdonik, S.: The 8 requirements of real-time stream processing. ACM SIGMOD Rec. **34**(4), 42–47 (2005)
48. Hulten, G., Spencer, L., Domingos, P.: Mining time-changing data streams. In: Proceedings of the Seventh ACM SIGKDD International Conference on Knowledge Discovery and Data Mining, pp. 97–106. ACM, New York (2001)
49. Zighed, D., Rakotomalala, R.: Graphes d'induction: apprentissage et data mining. Hermes Science Publications, Paris (2000)
50. Mehta, M., Agrawal, R., Rissanen, J.: SLIQ: a fast scalable classifier for data mining. In: Apers, P.M.G., Bouzeghoub, M., Gardarin, G. (eds.) EDBT 1996. LNCS, vol. 1057, pp. 18–34. Springer, Heidelberg (1996)
51. Shafer, J., Agrawal, R., Mehta, M.: SPRINT: a scalable parallel classifier for data mining. In: Proceedings of the International Conference on Very Large Data Bases, pp. 544–555 (1996)
52. Gehrke, J., Ramakrishnan, R., Ganti, V.: RainForest - a framework for fast decision tree construction of large datasets. Data Min. Knowl. Disc. **4**(2), 127–162 (2000)
53. Oates, T., Jensen, D.: The effects of training set size on decision tree complexity. In: ICML 1997: Proceedings of the Fourteenth International Conference on Machine Learning, pp. 254–262 (1997)
54. Hoeffding, W.: Probability inequalities for sums of bounded random variables. J. Am. Stat. Assoc. **58**, 13–30 (1963)
55. Matuszyk, P., Krempl, G., Spiliopoulou, M.: Correcting the usage of the hoeffding inequality in stream mining. In: Tucker, A., Höppner, F., Siebes, A., Swift, S. (eds.) IDA 2013. LNCS, vol. 8207, pp. 298–309. Springer, Heidelberg (2013)
56. Domingos, P., Hulten, G.: Mining high-speed data streams. In: Proceedings of the Sixth ACM SIGKDD International Conference on Knowledge Discovery and Data Mining, pp. 71–80. ACM, New York (2000)
57. Gama, J., Rocha, R., Medas, P.: Accurate decision trees for mining high-speed data streams. In: Proceedings of the Ninth ACM SIGKDD International Conference on Knowledge Discovery and Data Mining, pp. 523–528. ACM, New York (2003)
58. Ramos-Jiménez, G., del Campo-Avila, J., Morales-Bueno, R.: Incremental algorithm driven by error margins. In: Todorovski, L., Lavrač, N., Jantke, K.P. (eds.) DS 2006. LNCS (LNAI), vol. 4265, pp. 358–362. Springer, Heidelberg (2006)

59. del Campo-Avila, J., Ramos-Jiménez, G., Gama, J., Morales-Bueno, R.: Improving prediction accuracy of an incremental algorithm driven by error margins. Knowledge Discovery from Data Streams, 57 (2006)
60. Kirkby, R.: Improving hoeffding trees. Ph.D. thesis, University of Waikato (2008)
61. Kohavi, R., Kunz, C.: Option decision trees with majority votes. In: ICML 1997: Proceedings of the Fourteenth International Conference on Machine Learning, pp. 161–169. Morgan Kaufmann Publishers Inc., San Francisco (1997)
62. Breiman, L.: Bagging predictors. Mach. Learn. **24**(2), 123–140 (1996)
63. Robert, E., Freund, Y.: Boosting - Foundations and Algorithms. MIT Press, Cambridge (2012)
64. Wang, H., Fan, W., Yu, P.S., Han, J.: Mining concept-drifting data streams using ensemble classifiers. In: Proceedings of the Ninth ACM SIGKDD International Conference on Knowledge Discovery and Data Mining - KDD 2003, pp. 226–235. ACM Press, New York (2003)
65. Seidl, T., Assent, I., Kranen, P., Krieger, R., Herrmann, J.: Indexing density models for incremental learning and anytime classification on data streams. In: Proceedings of the 12th International Conference on Extending Database Technology: Advances in Database Technology, pp. 311–322. ACM (2009)
66. Rutkowski, L., Pietruczuk, L., Duda, P., Jaworski, M.: Decision trees for mining data streams based on the McDiarmid's bound. IEEE Trans. Knowl. Data Eng. **25**(6), 1272–1279 (2013)
67. Tsang, I., Kwok, J., Cheung, P.: Core vector machines: fast SVM training on very large data sets. J. Mach. Learn. Res. **6**(1), 363 (2006)
68. Dong, J.X., Krzyzak, A., Suen, C.Y.: Fast SVM training algorithm with decomposition on very large data sets. IEEE Trans. Pattern Anal. Mach. Intell. **27**(4), 603–618 (2005)
69. Usunier, N., Bordes, A., Bottou, L.: Guarantees for approximate incremental SVMs. In: Proceedings of the Thirteenth International Conference on Artificial Intelligence and Statistics, vol. 9, pp. 884–891 (2010)
70. Do, T., Nguyen, V., Poulet, F.: GPU-based parallel SVM algorithm. Jisuanji Kexue yu Tansuo **3**(4), 368–377 (2009)
71. Ferrer-Troyano, F., Aguilar-Ruiz, J.S., Riquelme, J.C.: Incremental rule learning based on example nearness from numerical data streams. In: Proceedings of the 2005 ACM Symposium on Applied Computing, p. 572. ACM (2005)
72. Ferrer-Troyano, F., Aguilar-Ruiz, J., Riquelme, J.: Data streams classification by incremental rule learning with parameterized generalization. In: Proceedings of the 2006 ACM Symposium on Applied Computing, p. 661. ACM (2006)
73. Gama, J.A., Kosina, P.: Learning decision rules from data streams. In: Proceedings of the Twenty-Second International Joint Conference on Artificial Intelligence, IJCAI 2011, vol. 2, pp. 1255–1260. AAAI Press (2011)
74. Gama, J., Pinto, C.: Discretization from data streams: applications to histograms and data mining. In: Proceedings of the 2006 ACM Symposium on Applied (2006)
75. Gibbons, P., Matias, Y., Poosala, V.: Fast incremental maintenance of approximate histograms. ACM Trans. Database **27**(3), 261–298 (2002)
76. Vitter, J.: Random sampling with a reservoir. ACM Trans. Math. Softw. **11**(1), 37–57 (1985)
77. Salperwyck, C., Lemaire, V., Hue, C.: Incremental weighted naive Bayes classifiers for data streams. Studies in Classification, Data Analysis, and Knowledge Organization. Springer, Heidelberg (2014)

78. Law, Y.-N., Zaniolo, C.: An adaptive nearest neighbor classification algorithm for data streams. In: Jorge, A.M., Torgo, L., Brazdil, P.B., Camacho, R., Gama, J. (eds.) PKDD 2005. LNCS (LNAI), vol. 3721, pp. 108–120. Springer, Heidelberg (2005)
79. Beringer, J., Hüllermeier, E.: Efficient instance-based learning on data streams. Intell. Data Anal. **11**(6), 627–650 (2007)
80. Shaker, A., Hüllermeier, E.: Iblstreams: a system for instance-based classification and regression on data streams. Evolving Syst. **3**(4), 235–249 (2012)
81. Cesa-Bianchi, N., Conconi, A., Gentile, C.: On the generalization ability of on-line learning algorithms. IEEE Trans. Inf. Theory **50**(9), 2050–2057 (2004)
82. Block, H.: The perceptron: a model for brain functioning. Rev. Mod. Phys. **34**, 123–135 (1962)
83. Novikoff, A.B.: On convergence proofs for perceptrons. In: Proceedings of the Symposium on the Mathematical Theory of Automata, vol. 12, pp. 615–622 (1963)
84. Cesa-Bianchi, N., Lugosi, G.: Prediction, Learning, and Games. Cambridge University Press, New York (2006)
85. Crammer, K., Kandola, J., Holloway, R., Singer, Y.: Online classification on a budget. In: Advances in Neural Information Processing Systems 16. MIT Press, Cambridge (2003)
86. Crammer, K., Dekel, O., Keshet, J., Shalev-Shwartz, S., Singer, Y.: Online passive-aggressive algorithms. J. Mach. Learn. Res. **7**, 551–585 (2006)
87. Shalev-Shwartz, S., Singer, Y., Srebro, N.: Pegasos: primal estimated sub-gradient solver for svm. In: Proceedings of the 24th International Conference on Machine Learning, ICML 2007, pp. 807–814. ACM, New York (2007)
88. Kivinen, J., Smola, A.J., Williamson, R.C.: Online learning with kernels. IEEE Trans. Sig. Process. **52**(8), 2165–2176 (2004)
89. Engel, Y., Mannor, S., Meir, R.: The kernel recursive least squares algorithm. IEEE Trans. Sig. Process. **52**, 2275–2285 (2003)
90. Csató, L., Opper, M.: Sparse on-line Gaussian processes. Neural Comput. **14**(3), 641–668 (2002)
91. Thompson, W.R.: On the likelihood that one unknown probability exceeds another in view of the evidence of two samples. Biometrika **25**, 285–294 (1933)
92. Bubeck, S., Cesa-Bianchi, N.: Regret analysis of stochastic and nonstochastic multi-armed bandit problems. Found. Trends Mach. Learn. **5**(1), 1–122 (2012)
93. Auer, P., Cesa-Bianchi, N., Freund, Y., Schapire, R.E.: The nonstochastic multi-armed bandit problem. SIAM J. Comput. **32**(1), 48–77 (2003)
94. Shawe-Taylor, J., Cristianini, N.: Kernel Methods for Pattern Analysis. Cambridge University Press, New York (2004)
95. Sutskever, I.: A simpler unified analysis of budget perceptrons. In: Proceedings of the 26th Annual International Conference on Machine Learning, ICML 2009, Montreal, Quebec, Canada, 14–18 June, pp. 985–992 (2009)
96. Dekel, O., Shalev-Shwartz, S., Singer, Y.: The forgetron: a kernel-based perceptron on a budget. SIAM J. Comput. **37**(5), 1342–1372 (2008)
97. Orabona, F., Keshet, J., Caputo, B.: The projectron: a bounded kernel-based perceptron. In: International Conference on Machine Learning (2008)
98. Bifet, A., Holmes, G., Pfahringer, B., Kirkby, R., Gavaldà, R.: New ensemble methods for evolving data streams. In: Proceedings of the 15th ACM SIGKDD International Conference on Knowledge Discovery and Data Mining - KDD 2009, p. 139 (2009)
99. Žliobaite, I.: Learning under concept drift: an overview. CoRR abs/1010.4784 (2010)

100. Lazarescu, M.M., Venkatesh, S., Bui, H.H.: Using multiple windows to track concept drift. Intell. Data Anal. **8**(1), 29–59 (2004)
101. Bifet, A., Gama, J., Pechenizkiy, M., Žliobaite, I.: Pakdd tutorial: Handling concept drift: Importance, challenges and solutions (2011)
102. Marsland, S.: Novelty detection in learning systems. Neural Comput. Surv. **3**, 157–195 (2003)
103. Faria, E.R., Goncalves, I.J.C.R., Gama, J., Carvalho, A.C.P.L.F.: Evaluation methodology for multiclass novelty detection algorithms. In: Brazilian Conference on Intelligent Systems, BRACIS 2013, Fortaleza, CE, Brazil, 19–24 October, pp. 19–25 (2013)
104. Gama, J., Medas, P., Castillo, G., Rodrigues, P.: Learning with drift detection. In: Bazzan, A.L.C., Labidi, S. (eds.) SBIA 2004. LNCS (LNAI), vol. 3171, pp. 286–295. Springer, Heidelberg (2004)
105. Baena-García, M., Del Campo-Ávila, J., Fidalgo, R., Bifet, A., Gavaldà, R., Morales-Bueno, R.: Early drift detection method. In: Fourth International Workshop on Knowledge Discovery from Data Streams, vol. 6, pp. 77–86 (2006)
106. Gama, J., Rodrigues, P.P., Sebastiao, R., Rodrigues, P.: Issues in evaluation of stream learning algorithms. In: Proceedings of the 15th ACM SIGKDD International Conference on Knowledge Discovery and Data Mining, pp. 329–338. ACM, New York (2009)
107. Page, E.: Continuous inspection schemes. Biometrika **41**(1–2), 100 (1954)
108. Mouss, H., Mouss, D., Mouss, N., Sefouhi, L.: Test of page-hinkley, an approach for fault detection in an agro-alimentary production system. In: 5th Asian Control Conference, vol. 2, pp. 815–818 (2004)
109. Bondu, A., Boullé, M.: A supervised approach for change detection in data streams (2011)
110. Boullé, M.: MODL: a Bayes optimal discretization method for continuous attributes. Mach. Learn. **65**(1), 131–165 (2006)
111. Minku, L., Yao, X.: DDD: a new ensemble approach for dealing with concept drift. IEEE Trans. Knowl. Data Eng. **24**, 619–633 (2012)
112. Widmer, G., Kubat, M.: Learning flexible concepts from streams of examples: FLORA2. In: Proceedings of the 10th European Conference on Artificial Intelligence. Number section 5, pp. 463–467. Wiley (1992)
113. Greenwald, M., Khanna, S.: Space-efficient online computation of quantile summaries. In: SIGMOD, pp. 58–66 (2001)
114. Breiman, L.: Bagging predictors. Mach. Learn. **24**(2), 123–140 (1996)
115. Street, W., Kim, Y.: A streaming ensemble algorithm (SEA) for large-scale classification. In: Proceedings of the Seventh ACM SIGKDD International Conference on Knowledge Discovery and Data Mining, pp. 377–382. ACM, New York (2001)
116. Kolter, J., Maloof, M.: Dynamic weighted majority: a new ensemble method for tracking concept drift. In: Proceedings of the Third International IEEE Conference on Data Mining, pp. 123–130 (2003)
117. Bifet, A., Gavalda, R.: Learning from time-changing data with adaptive windowing. In: SIAM International Conference on Data Mining, pp. 443–448 (2007)
118. Jaber, G.: An approach for online learning in the presence of concept changes. Ph.D. thesis, Université AgroParisTech (France) (2013)
119. Gama, J., Kosina, P.: Tracking recurring concepts with metalearners. In: Progress in Artificial Intelligence: 14th Portuguese Conference on Artificial Intelligence, p. 423 (2009)

120. Gomes, J.B., Menasalvas, E., Sousa, P.A.C.: Tracking recurrent concepts using context. In: Szczuka, M., Kryszkiewicz, M., Ramanna, S., Jensen, R., Hu, Q. (eds.) RSCTC 2010. LNCS, vol. 6086, pp. 168–177. Springer, Heidelberg (2010)

121. Salganicoff, M.: Tolerating concept and sampling shift in lazy learning using prediction error context switching. Artif. Intell. Rev. **11**(1), 133–155 (1997)

122. Cao, F., Ester, M., Qian, W., Zhou, A.: Density-based clustering over an evolving data stream with noise. In: 2006 SIAM Conference on Data Mining, pp. 328–339 (2006)

123. Quionero-Candela, J., Sugiyama, M., Schwaighofer, A., Lawrence, N.D.: Dataset Shift in Machine Learning. The MIT Press, Cambridge (2009)

124. Bifet, B., Gama, J., Gavalda, R., Krempl, G., Pechenizkiy, M., Pfahringer, B., Spiliopoulou, M., Žliobaite, I.: Advanced topics on data stream mining. Tutorial at the ECMLPKDD 2012 (2012)

125. Fawcett, T.: ROC graphs: notes and practical considerations for researchers. Mach. Learn. **31**, 1–38 (2004)

126. Bifet, A., Read, J., Žliobaité, I., Pfahringer, B., Holmes, G.: Pitfalls in benchmarking data stream classification and how to avoid them. In: Blockeel, H., Kersting, K., Nijssen, S., Železný, F. (eds.) ECML PKDD 2013, Part I. LNCS (LNAI), vol. 8188, pp. 465–479. Springer, Heidelberg (2013)

127. Žliobaité, I., Bifet, A., Read, J., Pfahringer, B., Holmes, G.: Evaluation methods and decision theory for classification of streaming data with temporal dependence. Mach. Learn. **98**, 455–482 (2015)

128. Dawid, A.: Present position and potential developments: some personal views: statistical theory: the prequential approach. J. Roy. Stat. Soc. Ser. A (General) **147**, 278–292 (1984)

129. Brzezinski, D., Stefanowski, J.: Prequential AUC for classifier evaluation and drift detection in evolving data streams. In: Proceedings of the Workshop New Frontiers in Mining Complex Patterns (NFMCP 2014) held in European Conference on Machine Learning (ECML) (2014)

130. Bifet, A.: Adaptive learning and mining for data streams and frequent patterns. Ph.D. thesis, Universitat Politécnica de Catalunya (2009)

131. Agrawal, R.: Database mining: a performance perspective. IEEE Trans. Knowl. Data Eng. **5**(6), 914–925 (1993)

132. Gama, J., Medas, P., Rodrigues, P.: Learning decision trees from dynamic data streams. J. Univ. Comput. Sci. **11**(8), 1353–1366 (2005)

133. Bifet, A., Kirkby, R.: Data stream mining a practical approach. J. Empirical Finance **8**(3), 325–342 (2009)

134. Minku, L.L., White, A.P., Yao, X.: The impact of diversity on online ensemble learning in the presence of concept drift. IEEE Trans. Knowl. Data Eng. **22**(5), 730–742 (2010)

135. Tavallaee, M., Bagheri, E., Lu, W., Ghorbani, A.A.: A detailed analysis of the kdd cup 99 data set. In: Proceedings of the Second IEEE International Conference on Computational Intelligence for Security and Defense Applications, CISDA 2009, pp. 53–58. IEEE Press, Piscataway (2009)

136. Katakis, I., Tsoumakas, G., Vlahavas, I.: Tracking recurring contexts using ensemble classifiers: an application to email filtering. Knowl. Inf. Syst. **22**(3), 371–391 (2010)

137. Witten, I.H., Frank, E.: Data Mining: Practical Machine Learning Tools and Techniques. Morgan Kaufmann Series in Data Management Systems, 2nd edn. Morgan Kaufmann, San Francisco (2005)

138. Žliobaité, I., Budka, M., Stahl, F.: Towards cost-sensitive adaptation: when is it worth updating your predictive model? Neurocomputing **150**, 240–249 (2014)
139. Bifet, A., Holmes, G., Pfahringer, B., Frank, E.: Fast perceptron decision tree learning from evolving data streams. In: Zaki, M.J., Yu, J.X., Ravindran, B., Pudi, V. (eds.) PAKDD 2010. LNCS, vol. 6119, pp. 299–310. Springer, Heidelberg (2010)
140. Littlestone, N., Warmuth, M.: The weighted majority algorithm. In: 30th Annual Symposium on Foundations of Computer Science, pp. 256–261 (1989)
141. Krempl, G., Žliobaite, I., Brzezinski, D., Hllermeier, E., Last, M., Lemaire, V., Noack, T., Shaker, A., Sievi, S., Spiliopoulou, M., Stefanowski, J.: Open challenges for data stream mining research. SIGKDD Explorations (Special Issue on Big Data) **16**, 1–10 (2014)

Knowledge Reuse: Survey of Existing Techniques and Classification Approach

Kurt Sandkuhl[✉]

Institute of Computer Science, University of Rostock, 18051 Rostock, Germany
Kurt.Sandkuhl@uni-rostock.de

Abstract. The importance of managing organizational knowledge for enterprises has been recognized since decades. The expectation is that systematic development and reuse of knowledge will help to improve the competitiveness of an enterprise. The paper investigates different approaches for knowledge reuse from computer science and business information systems. Starting from a description of different knowledge reuse perspectives and examples of knowledge reuse approaches from literature, the paper proposes a classification approach for knowledge reuse techniques. The criteria included in the approach are reuse technique, reuse situation, capacity of knowledge representation, addressee of knowledge, validation status, scope and phase of solution development. The main contributions of this paper is (1) a summary of the context for knowledge reuse from knowledge management and knowledge engineering (2) a set of criteria for comparing and categorizing knowledge reuse techniques derived from literature in the field, and (3) selected reuse techniques classified according to the criteria developed as initial validation of these classification scheme.

Keywords: Knowledge representation · Knowledge reuse · Module · Template · Pattern · Component · Knowledge classification

1 Introduction

The importance of managing organizational knowledge for enterprises has been recognized since decades. Enterprises and organizations establishing knowledge management structures perceive a number of advantages and benefits, like reducing problem-solving time, cost reduction of specific activities, faster delivery to market or better internal communication and increased staff participation [1]. The expectation is that systematic development and reuse of knowledge will help to improve the competitiveness of the enterprise under consideration [4]. Knowledge engineering [2] and enterprise knowledge modelling [3] contribute to this purpose by offering methods, tools and approaches for capturing knowledge in defined representations in order to support the entire lifecycle of organizational knowledge management.

Work presented in this paper focuses on a specific aspect of knowledge management and knowledge engineering: knowledge prepared for reuse. The paper investigates different approaches for knowledge reuse from computer science and business information systems. The literature in the field shows that different techniques have been investigated and applied for supporting knowledge reuse (cf. Sects. 2.3 and 3.1).

E. Zimányi and R.-D. Kutsche (Eds.): eBISS 2014, LNBIP 205, pp. 126–148, 2015.
DOI: 10.1007/978-3-319-17551-5_5

At a high abstraction level, reference models or frameworks are a recognized approach for capturing knowledge about solution design in general. The use of patterns is another accepted approach in this area and addresses a lower level of abstraction than frameworks. When it comes to reusing knowledge about procedures and role, process models and process patterns have been proposed, which often are part of the above mentioned frameworks. Furthermore, reuse techniques from different engineering disciplines, like components, modules or templates, have been adapted for knowledge management and knowledge engineering.

Experiences from industrial projects show that selecting the appropriate approach from this multitude of knowledge reuse techniques is difficult for practitioners (see, e.g. [5]. We argue that a classification of knowledge reuse approaches with clearly defined criteria motivated from industrial practice would help. As an important step into this direction we aim at developing such a classification approach. The main contributions of this paper is (1) a summary of the context for knowledge reuse from knowledge management and knowledge engineering (2) a set of criteria for comparing and categorizing knowledge reuse techniques derived from literature in the field, and (3) selected reuse techniques classified according to the criteria developed as initial validation of these classification scheme.

The remaining part of this paper is structured as follows: Sect. 2 will give background information to knowledge reuse including different interpretations of the term as such. Section 3 proposes criteria for comparing knowledge reuse approaches, i.e., a classification approach for knowledge reuse approaches. Section 4 illustrates the application of the classification approach by discussing two examples: ontology design patterns and the ITIL framework. In Sect. 5 validation of the classification approach by using it in different seminars is described. Section 6 summarizes the findings and discusses future work.

2 Knowledge Reuse

Our past research work included activities which contributed to both, the *organizational aspects* of knowledge management and knowledge representation, and the *technical aspects* of these fields. Regarding the organizational aspects, this included capturing organizational knowledge in enterprise models [7], enterprise knowledge models or organizational knowledge patterns [8]. The more technology-oriented work was in the field of ontology patterns [9] and ontology engineering practices [6]. During these research activities, we observed different perspectives on how to make knowledge reusable and how to define knowledge reuse as such. In order to illustrate these different perspectives, we discuss in Sect. 2.1 two of these perspectives: Nonaka and Takeuchi's SECI model [14] and Alan Newell's knowledge levels [12]. To some extent, these two perspectives also represent the different conceptual viewpoints on knowledge reuse from information systems and computer science. Furthermore, various literature studies performed as part of our research showed a multitude of techniques designed for facilitating knowledge reuse. Section 2.2 summarizes the results of this analysis. Moreover, Sect. 2.3 discusses the context of knowledge reuse in knowledge management and knowledge engineering. Section 2.3 aims at identifying related areas but does not have the ambition to provide an exhaustive description of the field.

2.1 Perspectives on Knowledge Reuse

Two important perspectives on knowledge reuse recognized by the scientific community originate from Nonaka and Takeuchi [14], and from Newell [12]. Both perspectives are summarized in this section in order to illustrate different positions to what knowledge reuse is.

An important dimension often discussed in knowledge management is the distinction between two types of knowledge, which is based on the work of Polanyi [11]: explicit knowledge and tacit knowledge. Explicit knowledge refers to knowledge that is codified, i.e., transmittable in a formal representation or language. Tacit knowledge is hard to formalize due to its personal quality of "simply knowing how to do something" in a specific context. Organizational knowledge includes both, tacit and explicit knowledge. In this context, the paper follows the opinion of Nonaka that an organization cannot create knowledge without individuals, i.e., at a fundamental level knowledge is created by individuals [10]. The organization supports individuals and provides a context for knowledge creation. Organizational knowledge creation includes processes that organizationally amplify the knowledge created by the individuals and crystallizes it as part of the knowledge network of the organization.

The SECI model of Nonaka and Takeuchi [14] describes the interrelation of tacit and explicit knowledge as well as steps required to create new knowledge and to transfer knowledge between actors. The basic structure of the SECI model is a matrix with two times two fields showing the possible transformation steps between tacit and explicit knowledge. One dimension in the matrix indicates the starting point of the knowledge transfer whereas the other marks the result dimension. Both are holding the characteristics tacit and explicit as displayed in Fig. 1.

The starting point of knowledge transfer between two actors is the direct transfer of tacit from one actor into tacit knowledge of the other actor. This is for instance done in a master-apprenticeship relationship when the master shows how to do it and the

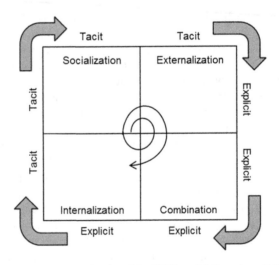

Fig. 1. Nonaka and Takeuchi's SECI model (based on [14])

apprentice imitates what the master showed him. This "socialization" does not require the knowledge to be available in explicit representation. The next phase, called explication, is happening when knowledge is made explicit to be stored or to be handed over. Nonaka and Takeuchi emphasize that this is not trivial, since knowledge is related to the inner "mental model" of the one carrying it. When this knowledge is explicit or stored in knowledge bases, the next step (combination) is to combine it with other explicit knowledge to create new knowledge, which again is in explicit form and can be stored. This could be done, e.g., by using reasoning or other approaches from artificial intelligence. The precondition for an actor to apply explicit knowledge is that it has to be internalized by the actor. This is the subject of the last phase. Since the four phases socialization, externalization, combination and internalization (SECI) are passed through repeatedly the model is also known as the spiral model. It nicely illustrates the processes needed to transfer knowledge between actors, i.e., to "reuse" knowledge between individuals.

Alan Newell's view on knowledge reuse and knowledge transfer is based on a different view what knowledge is [12, 13]. Knowledge is that which an observer ascribes to an intelligent agent (human or machine) that allows the observer to construe the agent's behavior as rational, i.e. behavior that allows the agent to achieve its perceived goals. Newell emphasizes that knowledge is an abstraction that cannot be written down. Data structures that we might use to encode knowledge in a computer knowledge base are not equivalent to the knowledge (the capacity for behavior) that those data structures represent. We are able to use data structures (symbols) to represent knowledge in a knowledge base, but those symbols cannot generate intelligent behavior - unless some process is applied to those symbols. This means we have to distinguish the symbols in a knowledge base (knowledge representation) from the knowledge (capacity for rational behavior) that the symbols can be used to generate.

Sharing and reuse of knowledge according to Newell requires specific preconditions. Knowledge bases have meaning only when they are processed by some interpreter - either by a computer program or by our own minds. We cannot share and reuse knowledge bases if we do not also share and reuse the inference engines (or mental processes) that bring our knowledge bases to life. Although we may speak of transferring "knowledge" from one site to another, we can at best transfer knowledge bases. We design our knowledge bases so that they can be processed to produce intelligent behavior. According to Newell, the area of problem solving is one of the application fields for sharing and reuse of knowledge. Sharing knowledge about problem solving requires a format for knowledge representation; a shared vocabulary; a conceptual model; and a process to be performed by the interpreter using the knowledge base (see also Sect. 3.3).

Comparing Newell's and Nonaka/Takeuchi's views, it can be concluded that these views are not conflicting. They share common ground but put an emphasis on different viewpoints. The common ground is that knowledge is more than what can be captured in knowledge representations or as explicit knowledge. Newell put an emphasis on the relation between knowledge base and the interpreter required to use the knowledge, whereas Nonaka and Takeuchi focus on the transition between tacit and explicit knowledge. Both works indicate many different options how to support reuse, e.g. starting either from Newell's knowledge levels or from the SECI phases, knowledge reuse techniques could be developed. Both works can also be used to derive potential classification criteria (see Sect. 3).

2.2 Knowledge Reuse Techniques

During our previous work on knowledge reuse techniques, we performed an analysis of related work for the organizational aspects of knowledge reuse and for the more technology-oriented aspects of this field. During this related work analysis, which is published in [15–17], many different techniques were discovered which either explicitly stated that they were designed for knowledge reuse or which due to their application context have to be positioned in this field. The purpose of this section is not to provide an exhaustive record of all existing techniques, but to illustrate the variety of knowledge reuse approaches.

Some examples of knowledge reuse techniques identified are:

- Semantic patterns [18].
- Knowledge patterns [28].
- Ontology modules [20].
- Ontology design patterns [29].
- Knowledge engineering macros [25].
- Task patterns [22].
- Information demand patterns [23].
- Knowledge architecture [30].
- Ontology architecture [21].
- Knowledge formalization patterns [19].
- Active Knowledge Architectures [3].
- Active Knowledge Models [3].
- Interaction Patterns [32].
- Knowledge prioritization templates [33].
- Knowledge Transformation Patterns [24].
- Workflow patterns [31].

Although many of the knowledge reuse approaches obviously have different intentions, others are not that clearly described and it is difficult to decide what the differences are and how to compare them. Furthermore, there is no obvious way to decide what the right approach for a specific purpose is.

2.3 Related Areas in Knowledge Management and Knowledge Engineering

Research in knowledge reuse is performed in various disciplines including economic sciences, psychology, engineering sciences or education sciences. In the scope of this paper with its focus on IT-related approaches and techniques, we will focus on computer science and business information systems. As mentioned earlier, the purpose of this paper is to investigate the field of knowledge reuse and propose a classification scheme. In this context it is important to understand related areas in order to position knowledge reuse with respect to related work. As discussed in Sects. 2.1 and 2.2, many approaches for knowledge reuse either originate from research on knowledge management or from knowledge engineering. Both areas aim at supporting the "lifecycle" of knowledge from inception to use and both include technological and methodical approaches.

The most relevant areas related to knowledge reuse are

- Knowledge management systems from an organizational perspective. These systems describe how to establish systematic knowledge management in an organization in terms of activities and organizational structures required. Well-known approaches sin this area are the "building block" model proposed by Probst et al. [34] and the SECI model already discussed in Sect. 2.1.
- Knowledge management systems from a technology perspective, i.e., IT-systems supporting organizational knowledge management. In this area, Maier et al.'s architecture proposal [35] for such systems and the differentiation between various knowledge services as components of this architecture is often applied.
- Knowledge representation techniques defined how explicit knowledge should be stored, e.g. as a knowledge base for different applications. Lamberts and Shanks [36] provides an overview to such techniques from computer science.
- Knowledge fusion addresses the question how to create new explicit knowledge from various knowledge sources, which often have different abstraction levels. Smirnov et al. [37] includes an overview to existing techniques in this field.
- Organizational situations for knowledge reuse were identified my Markus [38]. Knowledge about these situations supports the design of knowledge representation techniques and organizational practices.
- Evaluation of knowledge and knowledge management systems aims at deciding whether knowledge is useful for an organization and what the value is. Two selected approaches in this field are Delone and McLeans's IS success model [39] and Jennex and Olfman's approach [40].

Figure 2 illustrates the areas related to knowledge reuse along the two dimensions "lifecycle phase" and "organization/technology". In the middle part of Fig. 2 the lifecycle phases are illustrated from left to right by showing typical questions related to different phases of the use of knowledge: how to identify knowledge, how to store it, how to apply the knowledge or support its application, how to share it with other actors, how to evaluate the knowledge, and how to phase it out if it is no longer required? The upper part of Fig. 2 shows related areas which address primarily organizational aspects whereas the lower part of the figure is dedicated to the technology part. The rectangle in the center of the figure indicates the area where knowledge reuse is positioned.

3 Criteria for Comparing Knowledge Reuse Approaches

The different perspectives and the variety of approaches for knowledge reuse discussed in Sect. 2 form an important motivation for our work. To our knowledge there is no established approach for classification of knowledge reuse techniques covering

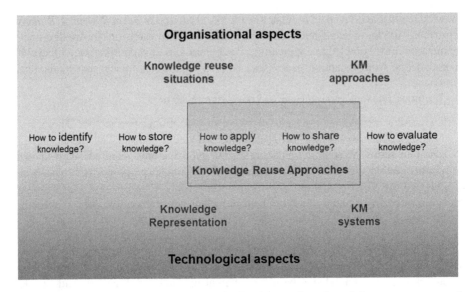

Fig. 2. Knowledge reuse in the context of related areas from knowledge management and knowledge engineering.

organisational and technology-focused approaches. Based on the analysis of literature in the field of knowledge management and knowledge engineering (see Sect. 2.3), based on own work in developing knowledge reuse techniques and methods and based on experiences in industrial application of knowledge reuse, this section proposes such a classification which consists of a number of criteria for comparing knowledge reuse approaches. The main purpose of these criteria is to support interested researchers and practitioners in navigating through the wealth of reuse approaches being published in the scientific literature.

The proposed criteria are discussed in the following sections:

- Reuse techniques (Sect. 3.1).
- Reuse situations (Sect. 3.2).
- Capacity of knowledge representation (Sect. 3.3).
- Addressee of knowledge (Sect. 3.4).
- Validation status (Sect. 3.5).
- Scope (Sect. 3.6).
- Phase of solution development (Sect. 3.7).

The selection of the above criteria was guided by the intention to represent different and complementary aspects of knowledge reuse from an organizational and a technological perspective. The criteria reuse situation, addressee of knowledge, validation status and phase of solution development are directed towards organizational aspects, whereas reuse technique, capacity of knowledge representation and scope address technological aspects. This set of criteria probably is not exhaustive, i.e., definition of additional criteria is possible and application of the criteria set is expected to show what additional ones are recommendable (see also Sect. 5).

All criteria aim at sorting knowledge reuse approaches into categories with respect to the criteria under consideration, i.e., all criteria contain a list of categories to be used. In case none of the given categories is applicable when using one of the criteria for a certain reuse approach, the additional category "other" should be used. Such a case would call for a discussion about an extension of the above classification scheme.

3.1 Reuse Techniques

In general, a technique denotes "practical method or art applied to some particular task[1]". In the context of knowledge reuse, we propose to consider techniques form computer science developed as means to facilitate reuse. Based on the literature search mentioned in Sect. 2.3, we identified four techniques frequently used in knowledge reuse:

- Module based techniques: a module is a self-contained component of a solution with defined interfaces hiding the actual implementation. The module has to be used "as is", i.e., without changing it, and often will be composed together other modules to a solution.
- Reference Architecture based techniques: architectures in general identify the main building blocks of a system with their interfaces and dependencies. Reference architectures are architectures reflecting the common building blocks for a defined domain which were agreed on by the stakeholders in that domain. Reference architecture can be considered as technique for knowledge reuse "in the large".
- Template based techniques: a template is a gauge to be used as a guide in making something accurately for a defined purpose. A template defines the structure but not the content; usually no behavioral aspects included.
- Pattern based techniques: a pattern provides solution principles (and how to implement them) for a recurring problem in a specific context by abstracting from actual application. A pattern exposes the core elements of the solution (structure and behavior) and consequences of using it. A pattern cannot be used as it is (unlike a module) but always has to be adapted for the purpose at hand.

The question to be answered for this criterion when classifying a knowledge reuse approach is: "What knowledge reuse technique is used by the knowledge reuse approach?".

3.2 Reuse Situations

The concept of reuse situations was proposed by Markus [38] as a way to characterize typical situations in organization where a demand of knowledge reuse arises. The reuse situations identified by Markus are:

- Shared work procedures: this situation exists when people work together on a team, either with the same roles or tasks or with different functions (cross-functional) and

---

[1] Source: www.vocabulary.com. Last access: October 13, 2014.

reuse work procedures earlier developed by themselves, i.e., they are producers of knowledge for their own later reuse.

- Shared work practitioners: when people in different organizational or geographical settings do similar work and share knowledge between each other in order to support their work. These people are not part of a team, but they have similar tasks. They are producers of knowledge for each other's reuse.
- Expertise-seeking novices: people with an occasional need for expert knowledge that they do not possess and do not need to acquire themselves because they need it rarely can be called expertise-seeking novices. Thus, this situation is given when a task has to be performed just once and the knowledge is required for this task only.
- Secondary knowledge miners: people who seek to answer new questions or develop new knowledge can be considered as "mining" for new knowledge. Through analysis of records or documentation produced by other people for different purposes, they aim to reuse knowledge.

The question to be answered for this criterion when classifying a knowledge reuse approach is: "For what knowledge reuse situation has the knowledge reuse approach been designed?".

3.3 Capacity of the Knowledge Representation

This criterion is based on Newell's work on knowledge levels (see Sect. 2.1). Capacity reflects how much of a problem solving task can be represented by the knowledge reuse approach under consideration, i.e., how "powerful" is the reuse approach when it comes to capturing all parts of the relevant knowledge. Regarding the capacity of the knowledge representation the following levels are distinguished:

- a knowledge representation format only defines how to represent the knowledge when explicating it. Often this includes syntax and semantics of records in knowledge bases, information structures in databases or languages to represent knowledge.
- reusable lexicon/shared vocabulary: in addition to the knowledge representation format, capturing knowledge usually requires the definition of what terms and concepts may be used for representing knowledge and what their meaning is.
- shared conceptual model: the relations between different concepts of the shared vocabulary has to be captured in order to represent knowledge. A knowledge reuse approach with the capacity to express a shared conceptual model usually allows for the definition of hierarchies, taxonomic relationships, classes of objects, characteristics of classes, etc.
- process reuse: according to Newell, the knowledge representation alone is not sufficient to capture knowledge but there has to be an interpreter (which can be a machine or the mental process of the human) using what is represented. If the knowledge reuse approach allows for representing the process to be performed by interpreter independent of knowledge representation format but dependent on the interpreter, it has the capacity of process reuse.

- Problem solving reuse: the highest capacity level is reached, if the process to be performed by an interpreter can be represented independent of the knowledge representation format and independent of the interpreter.

It should be noted that the above levels of knowledge representation are building on top of each other, i.e., "reusable lexicon/shared vocabulary" requires "knowledge representation format"; "shared conceptual model" requires "reusable lexicon/shared vocabulary" and "knowledge representation format"; etc.

The question to be answered for this criterion when classifying a knowledge reuse approach is: "What capacity does the knowledge representation underlying the reuse approach provide?".

3.4   Addressee of Knowledge

The addressee of knowledge captured by a reuse approach can be considered as the target group which is supposed to use the knowledge. Most reuse approaches capture knowledge which is meant to be used by an individual in her/his work context, but there are also knowledge reuse approaches suitable for organizational knowledge only. With this criterion, we aim to distinguish whether the knowledge reuse approach under consideration is meant for:

- an individual using the knowledge,
- a group of people, i.e., the knowledge is not meant or not to be used or not possible to use by an individual on its own, but it usually does not happen within an organizational context,
- an organisation,
- several organizations cooperating with each other (inter-organization).

The questioned to be answered for this criterion when classifying a knowledge reuse approach is: "What is the main target group of the knowledge provided by the reuse approach?".

3.5   Scope of the Knowledge

In the context of knowledge management, approaches for reusing knowledge often focus on specific perspectives of enterprise knowledge, like knowledge about processes or about products. The criteria "scope of knowledge" addresses this fact and aims at classifying knowledge reuse according to these perspectives. We propose to base this criterion on work from enterprise knowledge modeling. In general terms, enterprise modelling is addressing the systematic analysis and modelling of processes, organization structures, products structures, IT-systems or any other perspective relevant for the modelling purpose [41]. Enterprise knowledge modelling combines and extends approaches and techniques from enterprise modelling. The knowledge needed for performing a certain task in an enterprise or for acting in a certain role has to include the context of the individual, which requires including all relevant perspectives in the same model. Thus, an essential characteristic of knowledge models are "mutually reflective views of the different perspectives included in the model" [3]. As a best practice for capturing such mutually reflective views, the POPS* perspectives were proposed: the enterprise's processes (P), the organization structure (O), the product developed (P), the IT system used (S) and other aspects deemed relevant when modeling (*) [42].

Based on this best practice, the criterion is supposed to capture, what the main scope of the knowledge in the knowledge reuse approach is:

- Knowledge about a product. It should be noted that a product of an enterprise does not have to be a physical product but can be a service provided to a customer, i.e., it can be a "service product".
- Knowledge about an IT solution or artefact within the solution development process.
- Knowledge about a process.
- Knowledge about organizational structures.

The questioned to be answered for this criterion when classifying a knowledge reuse approach is: "What is the scope of the knowledge captured by the knowledge reuse approach?"

3.6 Phase of Solution Development

Both in knowledge management and in knowledge engineering, the introduction or development of systems or solutions to given problems happen in a systematic way, which is reflected in development phases. Many knowledge reuse approaches were designed for a specific development phase. The purpose of the criterion is to determine in which solution development phase the knowledge reuse approach under consideration is supposed to be useful or applicable. We distinguish between 7 traditional phases, which for example are reflected in software engineering approaches, like Boehm's spiral model [43]:

- Analysis.
- Specification.
- Design.
- Implementation.
- Verification and Validation.
- Operation.
- Maintenance.

The questioned to be answered for this criterion when classifying a knowledge reuse approach is: "For what phase of the solution development process situation has the knowledge reuse approach been designed?".

3.7 Validation Status of the Approach

The validation status of the reuse approach is considered an important criterion in order to, e.g., judge the suitability of the approach for use in industrial practice. Our assumption is that the more an approach has been validated in theory and practice, the more mature and useful it is. Among the many scientific approaches for validating research results, we base our proposal for judging the validation status on the work of Lincoln and Guba [44, p. 289 ff] on "naturalistic inquiry". On the one hand, we distinguish between theoretical and practical validation. Theoretical validation means assessing an approach within the theories of the domain the approach is part of or supposed to contribute to. In the context of knowledge reuse, this means to assess the soundness, feasibility, consistency within the body of knowledge in, for instance, knowledge management and knowledge engineering. Practical validation encompasses all kinds of application of the approach for validation purposes, which requires defined procedures and documenting results. This could be simple lab examples illustrating the approach, controlled experiments in a lab setting, application in industrial cases, etc.

On the other hand, we consider the context of validation and distinguish between validation by the developers of the approach in their internal environment, validation by the developers outside the internal environment, and validation by other actors than the developers. Combining these two perspectives leads to a two by three matrix, which is depicted in Table 1. The cells of this table show typical ways of validation for the different combinations of the two perspectives.

Table 1. Validation steps according to Lincoln and Guba

	Theory	Practice
Internal, development team	Validation against state of research, internal consistency checks	Prototype implementation for checking feasibility, test in lab environment
External, in validation context	Peer-review of publications describing approach and concepts, comparison to known best practices of the domain	Case studies with application partners using the artifacts for evaluation purposes
		Application of the developed artifacts in cooperation/under instruction from developers
External, in application context	Development of extensions or enhancements of the concepts and approaches by external actors	Use of the artifacts developed (e.g. algorithms, methods, software components) for solutions
	Application of the artifacts for creation of new theoretical knowledge	
	Comparison with related approaches	

Using the above matrix, information about the knowledge reuse approaches has to be used to determine where to position the validation status for the reuse approach in the matrix. Usually, validation starts on the "internal, development team" level with validation in theory followed by validation in practice, and proceeds "downward" in the matrix with alternating theory and practice validation to "external, in application context". Thus, the highest validation status would be reached if all cells in the matrix were covered.

The questioned to be answered for this criterion when classifying a knowledge reuse approach is: "What validation status does the knowledge reuse approach have?".

4 Example of Applying the Classification Approach

Section 3 defined a set of criteria for classifying knowledge reuse approaches. This section presents two knowledge reuse approaches in more detail and shows the use of the criteria for classifying these examples: ontology design patterns (Sect. 4.1) and the reference model ITIL.

4.1 Ontology Design Patterns

In a computer science context, ontologies usually are defined as explicit specifications of a shared conceptualization [45]. Due to the increasing use of ontologies in industrial applications, ontology design, ontology engineering and ontology evaluation have become a major concern. The aim is to efficiently produce high quality ontologies as a basis for semantic web applications or enterprise knowledge management.

Despite quite a few well-defined ontology construction methods and a number of reusable ontologies offered on the Internet, efficient ontology development continues to be a challenge, since this still requires a lot of experience and knowledge of the underlying logical theory.

Ontology Design Patterns (ODP) are considered a promising contribution to this challenge. In 2005, the term ontology design pattern in its current interpretation was mentioned by Gangemi [46] and introduced by Blomqvist and Sandkuhl [9]. Blomqvist defines the term as "a set of ontological elements, structures or construction principles that solve a clearly defined particular modelling problem" [29]. Ontology design patterns are considered as encodings of best practices, which help to reduce the need for extensive experience when developing ontologies, i.e., the well-defined solutions encoded in the patterns can be exploited by less experienced engineers when creating ontologies.

Different types of ODP are under investigation, which are discussed in [47] regarding their differences and the terminology used. The two types of ODP probably receiving most attention are logical and content ODP. Logical ODP focus only on the logical structure of the representation, i.e., this pattern type is targeting aspects of language expressivity, common problems and misconceptions. Content ODP often are instantiations of logical ODP offering actual modelling solutions. Due to the fact that these solutions contain actual classes, properties, and axioms, content ODP are considered by many researchers as domain-dependent, even though the domain might be considering general issues like 'events' or 'situations'. Platforms offering ODP currently include the ODP wiki portal[2] initiated by the NeOn-project[3] and the logical ODPs maintained by the University of Manchester[4].

---

[2] http://ontologydesignpatterns.org.

[3] EU-FP7 funded IP that ended in February 2010 - http://www.neon-project.org.

[4] http://www.gong.manchester.ac.uk/odp/html/index.html.

An example of a content ODP is the "Bag" pattern, which initially was proposed by Blomqvist [29] and is available in the ODP portal catalogue.[5] The pattern specifies how to represent a collection of items (a "bag") when this collection can have multiple copies of each item. Figure 3 illustrates the OWL building block of the pattern. The pattern catalogue provides additional information according to a pattern template, including name, intent, domain addressed, author, competency questions (showing the requirements covered), scenarios, consequences, and a link to a reusable OWL model representing the solution proposed by the pattern.

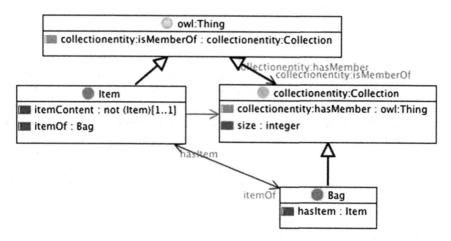

Fig. 3. The "Bag" content ODP, illustrated using OWL

When classifying ODP with the approach presented in Sect. 3, it is obvious that the technique used in ontology design patterns is "pattern based technique". Since ODP are meant for sharing experiences between ontology engineers, the reuse situation is "shared work practitioners". ODP are meant to capture not only a conceptual model, but all information for problem solving, which shows the capacity of the approach. ODP will primarily be used by individuals, which makes individuals the primary addressee of the knowledge in ODPs. The scope of the knowledge is an IT solution, as ODP are used to construct IT solutions. The term ODP indicates that the development phase addressed is "design". As ODP are used in application context by engineers external to the development teams, the validation status of the approach is "external, in application context". The result of the classification is shown in Table 2.

4.2 ITIL Framework

The Information Technology Infrastructure Library (ITIL) [48] is becoming more and more important for IT service management in the industrial and administrative sector. ITIL basically consists of a set of concepts and practices for Information Technology

---

[5] http://ontologydesignpatterns.org/wiki/Submissions:Bag.

Table 2. Classification of ODP

Criterion	Classification of ODP
Reuse technique	Pattern
Reuse situation	Shared work practitioners
Capacity of knowledge reuse approach	Problem solving reuse
Addressee of knowledge	Individual
Scope of the knowledge	Knowledge about an IT solution
Phase of solution development	Design
Validation status of the approach	External, in application context

Services Management (ITSM), Information Technology development and IT operations. IT Service Management in general is concerned with delivering and supporting IT services that are appropriate to the business requirements of an organization. This is supposed to improve efficiency and effectiveness and reduces the risks of managing IT services. ITIL gives detailed descriptions of a number of important IT practices and provides comprehensive checklists, tasks and procedures that any IT organization can tailor to its needs.

The ITIL reference model has been developed by the government agency Office of Government Commerce (OGC) in the UK. Version 3 of the reference model (ITIL V3) from the year 2007 was published in a series of six books (Introduction to ITIL Service Management, Service Strategy, Service Design, Service Transition, Service Operation und Continual Service Improvement). This series is called the ITIL core publication and is under continuous development. The non-proprietary license of the ITIL reference model allows for the free distribution of the core ideas and is published by the UK-based privatized government publishing house 'The Stationary Office' (TSO) [48, S. XXXIX].

The volume Service Operation, to take just one example, consists of the management practices of daily service operation. The service operation lifecycle phase is on the one hand responsible for the coordination and execution of the activities and processes. On the other hand this phase is also responsible for operating the technology needed for the delivery of the IT services as defined in the service level agreements. Compliance of the contracted service levels is a top priority. This includes the availability of the IT service so that the customer can use the service. This also includes the quality of service for example in terms of performance in order to ensure both the usability and utility of the IT service following the requirements of the customer.

Service operation is also responsible for the management of the technology that is necessary for the service delivery and service maintenance. Supervising the performance by monitoring the IT service as well as processing events for example by analyzing log file data plays an important role. Out of the information obtained conclusions have to be made and applied within the incident and problem solution process. A continuous improvement of the IT services is based on the results of the monitoring and evaluation activities of service operation [49].

When classifying ITIL with the approach presented in Sect. 3, the technique used for ITIL is "reference model based technique". Since ITIL is meant for sharing experiences between people implementing IT service management, the reuse situation is "shared work practitioners". ITIL primarily captures only recommendations and structures for the problem at hand, which have to be adapted to companies' needs. This shows that the capacity of the approach is "conceptual model". ITIL will primarily be used on organizational level, which makes organizations the primary addressee of the knowledge in ODPs. The scope of the knowledge includes process and organization structure, since ITIL covers both aspects. The reference model can be used for analyzing existing situations in an enterprise and designing the future status. This indicates that the development phase addressed is "analysis" and "design". As ITIL is used in application context by many companies, which were not involved in its development, the validation status of the approach is "external, in application context". The result of the classification is shown in Table 3.

Table 3. Classification of ITIL

Criterion	Classification of ODP
Reuse technique	Reference model
Reuse situation	Shared work practitioners
Capacity of knowledge reuse approach	Conceptual model
Addressee of knowledge	Organization
Scope of the knowledge	process and organization structures
Phase of solution development	Analysis, design
Validation status of the approach	External, in application context

5   Validation of the Classification Approach

The validation of the classification approach so far included validation in theory and practice in both, internal and external contexts. Parts of the internal validation were internal consistency checks and comparison against the state of the art, and application of the approach for classifying two knowledge reuse approaches (see Sect. 4). This section will primarily focus on external validation.

External validation requires that the approach under consideration is applied and evaluated by actors which were not involved in developing it. This kind of evaluation was performed at three different occasions with different student groups in different educational settings which all can be considered as seminars:

- The first seminar happened in spring 2013 in a PhD course on Knowledge Engineering with Semantic Web Technologies at Linköping University, Sweden. Participants were 12 PhD students, most of them with a background in computer science (9) and some with an education in information engineering (2) or logistics (1).
- Seminar 2 was organized in autumn 2013 in a master course on Knowledge Representation at Rostock University, Germany. Participants were 8 master students, all of them with a Bachelor in Business Information Systems.

Table 4. Papers on knowledge reuse techniques selected for the seminars

No.	Paper
1	Staab, S., Erdmann, M. and Maedche, A. (2001) Engineering Ontologies using Semantic Patterns. In D. O'Leary and A. Preece (eds.) Proceedings of the IJCAI-01 Workshop on E-business & The Intelligent Web, Seattle, 2001
2	Puppe, F. (2000) Knowledge Formalization Patterns. In Proceedings of PKAW 2000, Sydney, Australia, 2000, 2000
3	Doran, P., Tamma, V. and Iannone, L. (2007) Ontology module extraction for ontology reuse: an ontology engineering perspective. In Proceedings of the sixteenth ACM conference on Conference on information and knowledge management (CIKM '07). ACM, New York, NY, USA, 61–70
4	Lee, J., Chae, H., Kim, K. and Kim, C. H. (2006) An Ontology Architecture for Integration of Ontologies. The Semantic Web – ASWC 2006. R. Mizoguchi, Z. Shi and F. Giunchiglia, Springer Berlin/Heidelberg. 4185: 205–211
5	Sandkuhl, K. (2010) Capturing Product Development Knowledge with Task Patterns: Evaluation of Economic Effects. Quarterly Journal of Control & Cybernetics, Issue 1, 2010. Systems Research Institute, Polish Academy of Sciences
6	Sandkuhl, K. (2011) Improving Engineering Change Management with Information Demand Patterns. Pre-Proceedings of the IFIP WG 5.1 8th International Conference on Product Lifecycle Management, Eindhoven (NL). Inderscience Enterprises Ltd
7	Svátek, V. (2004) Design Patterns for Semantic Web Ontologies: Motivation and Discussion. Business Information Systems, Proceedings of BIS 2004, Poznań, Poland
8	Vrandecic, D. (2005) Explicit knowledge engineering patterns with macros. Proceedings of the Ontology Patterns for the Semantic Web Workshop at the ISWC 2005
9	Gracia, J., Liem, J., Lozano, E., Corcho, O., Trna, M., Gómez-Pérez, A., & Bredeweg, B. (2010). Semantic techniques for enabling knowledge reuse in conceptual modelling. *The Semantic Web–ISWC 2010*, 82–97
10	Gaeta, M., Orciuoli, F., Paolozzi, S., & Salerno, S. (2011). Ontology extraction for knowledge reuse: The e-learning perspective. *Systems, Man and Cybernetics, Part A: Systems and Humans, IEEE Transactions on, 41*(4), 798–809

- Seminar 3 took place as part of the eBISS[6] summer school in summer 2014 at Berlin University of Technology. Participants were more than 40 PhD students and master students from different European universities and scholarship holders from many non-European countries.

All three seminars followed the same set-up and work process:

- a 4 hour time frame was arranged, which consisted of 1,5 hours lecture, 1,5 hours group work and one hour discussion of the group work results
- The lecture introduced the term knowledge reuse, related areas from knowledge management and knowledge engineering, examples for knowledge reuse techniques

---

[6] eBISS = 4th European Business Intelligence Summer School, Berlin, July 2014.

and the classification approach, i.e., the lecture basically had the same content as Sects. 1–4 of this paper,

- The group work part had the task for each group to use the classification approach for investigating and classifying a knowledge reuse technique. Each of the techniques to be classified was documented in a scientific publication. The participants formed groups of two or three students and selected themselves which technique to work on. No group was allowed to take the same paper as another group. The techniques selected for the group work are listed in Table 4 and include [26, 27],

- The discussion of the group work results consisted for each group of a short description of the publication analyzed and the classification reached. This classification was discussed with the other seminar participants. Furthermore, each group had the explicit task to reflect on suitability of the different classification criteria for the given purpose, on the applicability of the criteria, i.e., are the criteria described in a way which allows to clearly distinguish the different categories, and whether criteria or aspects for classification were missing or superfluous. In seminar 1 and 2, the participants had to summarize their results in a short report; in seminar 3, the results were captured during discussion.

The results of the group work in the three seminars are summarized in Table 5 regarding the classifications which were detected for each technique. Due to the different number of participants and due to the possibility for each group to select their publication to evaluate, not all papers were classified in each seminar. Furthermore, in seminar 3 each paper was classified by two different groups simultaneously. Regarding the suitability of the classification criteria, the participants of the seminar confirmed that it was possible to use the criteria for analyzing the characteristics of knowledge reuse techniques and to classify the techniques described in the provided papers. The classification results discussed above and shown in Table 5 support this impression. For the applicability, most criteria were perceived as sufficiently clear defined and applicable in practical use: reuse technique, reuse situation, capacity of knowledge representation, validation status, and phase of solution development. The criterion which received criticism only in seminar 3 was the scope of the knowledge. The participants of this seminar expressed that the product and process perspective were not sufficiently distinguishable. Additional explanations by the teacher during the seminar solved this issue. Seminar 1 and 2 did not raise this question, probably because they both had a solid education in enterprise modeling which emphasizes the distinction between product and process knowledge. The implication for the classification approach is that the aspect should be more clearly explained and illustrated with examples.

Criticism from all three seminars was expressed regarding the target group criterion "addressee". A knowledge representation technique suitable for an individual in his daily work also will be of use for the organization this individual is working with. On the other side, knowledge meant for the organization will usually be applied by the individuals in the organization. In order to: If the knowledge reuse technique cannot be applied by an individual alone but need organizational structures, like roles or processes, than the technique is to be classified as meant for organizations, not for individuals.

Table 5. Summary of the classification results from all seminars for the papers listed in Table 4

Paper no.	Reuse Technique	Reuse Situation	Capacity	Addressee	Scope	Phase of Development	Validation status of Approach
1	1, 2: pattern; 3: template or pattern	1, 2: n/a; 3: shared work practitioners	all: conceptual model	all: group	all: process	1, 2: design; 3: design + validation	all: external, in validation context
2	1: pattern; 2: n/a; 3: template/patt.	1, 2: n/a; 3: expertise-seeking novices	1, 3: concept. model; 2: n/a	1, 3: group; 2: n/a	1, 3: process + IT-artefact; 2: n/a	1,3: design; 2: n/a	1, 3: external, in validation context; 2: n/a
3	1,2: module; 3: n/a	1, 2: shared work practitioners; 3: n/a	1, 2: concept. model; 3: n/a	1, 2: individual; 3: n/a	1, 2: IT-artefact; 3: n/a	1, 2: design; 3: n/a	1, 2: external, in validation context; 3: n/a
4	1: reference model; 2, 3: n/a	1: shared work practitioners; 2, 3: n/a	1: concept-ual model; 2, 3: n/a	1: individual; 2, 3: n/a	1: IT-artefact; 2, 3: n/a	1: analysis + design; 2, 3: n/a	1: internal, in validation context; 2, 3: n/a
5	1: pattern; 2: n/a; 3: template	1, 3: shared work procedures; 2: n/a	1, 3: problem solving; 2: n/a	1,3: organization; 2: n/a	1,3: process, org., product, IT; 2: n/a	1, 3: design; 2: n/a	1,3: external, in application context; 2: n/a
6	1,2: pattern; 3: n/a	1, 2: expertise-seeking novices; 3: n/a	1, 2: concept. model; 3: n/a	1, 2: individual; 3: n/a	1, 2: organization; 3: n/a	1, 2: analysis, design; 3: n/a	1,2: external, in application context; 3: n/a
7	2: pattern; 1, 3: n/a	2: shared work practitioners; 1, 3: n/a	2: conceptual model; 1, 3: n/a	2: individual; 1, 3: n/a	2: IT-artefact; 1, 3: n/a	2: analysis, design; 1, 3: n/a	2: internal, validation context; 1,3: n/a
8	1,3: pattern; 2: n/a	1,3: shared work practitioners; 2: n/a	1,3: process reuse; 2: n/a	1,3: individual; 2: n/a	1,3: IT-artefact; 2: n/a	1,3: implementation; 2: n/a	1,3: external, validation context; 2: n/a
9	1,2: n/a; 3: pattern	1,2: n/a; 3: secondary knowledge miners	1,2: n/a; 3: conceptual model	1,2: n/a; 3: individual	1,2: n/a; 3: process	1,2: n/a; 3: design	1,2: n/a; 3: internal, validation in practice
10	1, 2: n/a; 3: reference model	1, 2: n/a; 3: second. Know. miners	1, 2: n/a; 3: process reuse	1, 2: n/a; 3: individual/ group	1, 2: n/a; 3: process, product	1, 2: n/a; 3: design, implementation	1,2: n/a; 3: external, validation context

6 Summary and Future Work

Starting from a description of different knowledge reuse perspectives and examples of knowledge reuse approaches from literature, the paper presented a classification approach for knowledge reuse techniques. The criteria included in the approach are reuse technique, reuse situation, capacity of knowledge representation, addressee of knowledge, validation status, scope and phase of solution development. Suitability and applicability of the approach are evaluated by applying them on different knowledge reuse approaches which by intention were selected from different areas of computer science and business information systems (ontology design patterns and ITIL reference model) and by using them as part of an assignment in different seminars in university education on PhD and master student level. The classification approach was perceived applicable, suitable and useful for the intended purpose.

One part of future work will be to perform various refinement and improvement activities of the classification approach:

- The criteria receiving criticism during application in the seminars need improvement.
- The categorization for each individual criterion should be checked one more time for completeness. For this purpose an extensive literature study will be performed.
- The way of how to perform the classification for a given knowledge reuse approach should be described in more detail as a guideline.

Another important part of the future work will be to revisit the initial motivation for developing the classification: to support practitioners in finding and selecting the right knowledge reuse approach for a given problem or application scenario. For this purpose, much information included in the classification is supposed to be useful and required, like the reuse situation and the technique. However, it will be crucial to better understand the drivers and frame conditions of knowledge reuse in organizations. Probably, typical motivations like automation for higher efficiency or standardization as means to raise quality will not be sufficient. An analysis of literature on knowledge reuse cases and new case studies in this field is expected to provide first insights.

Acknowledgement. We would like to thank the anonymous reviewers for their valuable comments which helped to improve the paper.

References

1. Alavi, M., Leidner, D.E.: Knowledge management systems: issues, challenges, and benefits. Commun. AIS **1**(2es), 1 (1999)
2. Studer, R., Benjamins, V.R., Fensel, D.: Knowledge engineering: principles and methods. Data Knowl. Eng. **25**(1), 161–197 (1998)
3. Lillehagen, F., Krogstie, J.: Active Knowledge Modelling of Enterprises. Springer, Heidelberg (2009). ISBN 978-3-540-79415-8
4. Dalkir, K.: Knowledge management in theory and practice. Routledge, London (2013)

5. Sandkuhl, K., Lillehagen, F.: The early phases of enterprise knowledge modelling: practices and experiences from scaffolding and scoping. In: Stirna, J., Persson, A. (eds.) The Practice of Enterprise Modeling. Lecture Notes in Business Information Processing, vol. 15, pp. 1–14. Springer, Heidelberg (2008)

6. Sandkuhl, K., Öhgren, A., Smirnov, A., Shilov, N., Kashevnik, A.: Ontology construction in practice - experiences and recommendations from industrial cases. In: 9th International Conference on Enterprise Information Systems, Funchal, Madeira, Portugal, 12–16, June 2007

7. Sandkuhl, K., Stirna, J., Persson, A., Wißotzki, M.: Enterprise Modeling: Tackling Business Challenges with the 4EM Method. Springer (2014)

8. Sandkuhl, K.: Organizational knowledge patterns - definition and characteristics. In: Proceedings of the International Conference on Knowledge Management and Information Sharing, pp. 230–235. SceTePress, Paris, France, 26–29 October 2011. ISBN: 978-989-8425-81-2

9. Blomqvist E., Sandkuhl, K.: Patterns in ontology engineering – classification of ontology patterns. In: Proceedings of the 7th International Conference on Enterprise Information Systems, Miami, USA, May 2005

10. Nonaka, I.: A dynamic theory of organizational knowledge creation. Organ. Sci. 5(1), 14–37 (1994)

11. Polanyi, M.: Personal Knowledge. The University of Chicago Press, Chicago (1958)

12. Newell, A.: The knowledge level. Artificial Intelligence 18, 87 (1982). (Association for the Advancement of Artificial Intelligence)

13. Musen, M.A.: Dimensions of knowledge sharing and reuse. Comput. Biomed. Res. 25, 435–467 (1992). Academic Press

14. Nonaka, I., Takeuchi, H.: The Knowledge-Creating Company: How Japanese Companies Create the Dynamics of Innovation. Oxford University Press, New York (1995)

15. Wißotzki, M., Sandkuhl, K., Dunkel, A., Viktorya Christina, L.: State of research in reuse of enterprise models: systematic literature analysis of CAISE, EMMSAD, ICIS and INCOM. In: Isaías, P., Baptista Nunes, M., Powell, P. (eds.) Proceedings IADIS, Information Systems 2012, pp. 82–90. IADIS Press, Berlin (2012). ISBN: 978-972-8939-68-7

16. Hammer, K., Sandkuhl, K.: The state of ontology pattern research: a systematic review of ISWC, ESWC and ASWC 2005–2009. In: International Workshop on Ontology Patterns WOP 2010 at 9th International Semantic Web Conference, Shanghai, 8–10 November 2010. CEUR Workshop Proceedings, vol. 671 (2010). http://ceur-ws.org. ISSN 1613-0073

17. Sandkuhl, K.: Organizational knowledge patterns: foundation and application example. In: Villa-Vargas, L., Sheremetov, L., Haasis, H.D. (eds.) Operations Research and Data Mining ORADM 2012 Workshop Proceedings. Cancun, Mexico. Mexican Petroleum Institute, Mexico City, March 2012. 978-607-414-284-6

18. Staab, S., Erdmann, M., Maedche, A.: Engineering ontologies using semantic patterns. In: O'Leary, D., Preece, A. (eds.) Proceedings of the IJCAI-01 Workshop on E-business and the Intelligent Web, Seattle (2001)

19. Puppe, F.: Knowledge formalization patterns. In: Proceedings of PKAW 2000, Sydney, Australia (2000)

20. Doran, P., Tamma, V., Iannone, L.: Ontology module extraction for ontology reuse: an ontology engineering perspective. In Proceedings of the sixteenth ACM conference on Conference on information and knowledge management (CIKM 2007), pp. 61-70. ACM, New York, NY, USA (2007)

21. Lee, J., Chae, H., Kim, K., Kim, C.-H.: An ontology architecture for integration of ontologies. In: Mizoguchi, R., Shi, Z., Giunchiglia, F. (eds.) ASWC 2006. LNCS, vol. 4185, pp. 205–211. Springer, Heidelberg (2006)

22. Sandkuhl, K.: Capturing product development knowledge with task patterns: evaluation of economic effects. Q. J. Control Cybern. **39**(1), 259–273 (2010). Systems Research Institute, Polish Academy of Sciences

23. Sandkuhl, K.: Improving engineering change management with information demand patterns. In: Pre-Proceedings of the IFIP WG 5.1 8th International Conference on Product Lifecycle Management. Inderscience Enterprises Ltd, Eindhoven (NL) (2011)

24. Svátek, V.: Design patterns for semantic web ontologies: motivation and discussion. In: Proceedings of Business Information Systems, BIS 2004, Poznań, Poland (2004)

25. Vrandecic, D.: Explicit knowledge engineering patterns with macros. In: Proceedings of the Ontology Patterns for the Semantic Web Workshop at the ISWC (2005)

26. Gracia, J., Liem, J., Lozano, E., Corcho, O., Trna, M., Gómez-Pérez, A., Bredeweg, B.: Semantic techniques for enabling knowledge reuse in conceptual modelling. In: The Semantic Web–ISWC 2010, pp. 82–97 (2010)

27. Gaeta, M., Orciuoli, F., Paolozzi, S., Salerno, S.: Ontology extraction for knowledge reuse: the e-learning perspective. IEEE Trans. Syst. Man Cybern. Part A Syst. Hum. **41**(4), 798–809 (2011)

28. Clark, P., Thompson, J., Porter, B.: Knowledge patterns. In: Cohn, A.G., Giunchiglia, F., Selman, B.(eds.) KR2000: Principles of Knowledge Representation and Reasoning. Morgan Kaufman, San Francisco (2000)

29. Blomqvist, E.: Semi-automatic ontology construction based on patterns. Ph.D. thesis, Department of Computer and Information Science, Linköping University (2009)

30. Sandkuhl, K.: Pattern use in knowledge architectures:an example from information logistics. In: Kobyliński, K., Sobczak, A. (eds.) Perspectives in Business Informatics Research. Lecture Notes in Business Information Processing, vol. 158, pp. 91–103. Springer, Heidelberg (2013). ISBN 978-3-642-40822-9

31. Van der Aalst, W.M.P., ter Hofstede, A.H.M., Kiepuszewski, B., Barros, A.P.: Workflow Patterns. Distrib. Parallel Databases **14**, 5–51 (2003). (Kluwer)

32. Schümmer, T., Lukosch, S.: Patterns for Computer-Mediated Interaction. Wiley, Chichester (2007). ISBN 978-0-470-02561-1

33. Levy, M.: Knowledge retention: minimizing organizational business loss. J. Knowl. Manage. **15**(4), 582–600 (2011)

34. Probst, G., Raub, S., Romhardt, K.: Managing Knowledge – Building Blocks for Success. Wiley, Chichester (2000)

35. Maier, R., Hädrich, T., Peinl, R.: Enterprise Knowledge Infrastructures, 2nd edn. Springer, Heidelberg (2010)

36. Lamberts, K., Shanks, D. (eds.): Knowledge Concepts and Categories. Psychology Press, Cambridge (2013)

37. Smirnov, A.; Sandkuhl, K., Shilov, N.: Multilevel self-organisation and context-based knowledge fusion for business model adaptability in cyber-physical systems. In: Bakhtadze, N. (ed.) Manufacturing Modelling, Management, and Control. With assistance of Natalia Bakhtadze, Kirill Chernyshov, Alexandre Dolgui, Vladimir Lototsky. Saint Petersburg, Russia, vol. #7, pp. 609–613. Elsevier, IFAC (IFAC proceedings volumes) 19 June 2013

38. Lynne Markus, M.: Toward a Theory of knowledge reuse: types of knowledge reuse situations and factors in reuse success. J. Manage. Inf. Syst. **18**(1), 57–93 (2001)

39. Delone, W.H.: The delone and mclean model of information systems success: a ten-year update. J. Manage. Inf. Syst. **19**(4), 9–30 (2003)

40. Jennex, M.E., Olfman, L.: A model of knowledge management success. Int. J. Knowl. Manage. (IJKM) **2**(3), 51–68 (2006)

41. Vernadat, F.B.: Enterprise Modelling and Integration. Chapman & Hall, London (1996)

42. Lillehagen, F.: The foundations of AKM technology. In: Proceedings 10th International Conference on Concurrent Engineering (CE) Conference, Madeira, Portugal (2003)
43. Boehm, B.W.: A spiral model of software development and enhancement. IEEE Comput. **21** (5), 61–72 (1988)
44. Lincoln, Y., Guba, E.: Naturalistic Inquiry. Sage Publications, Beverly Hills (1985)
45. Gruber, T.: A translation approach to portable ontology specifications. Knowl. Acquis. **5**, 199–220 (1993)
46. Gangemi, A.: Ontology design patterns for semantic web content. In: Gil, Y., Motta, E., Richard Benjamins, V., Musen, M.A. (eds.) TISWC 2005. LNCS, vol. 3729, pp. 262–276. Springer, Heidelberg (2005)
47. Gangemi, A., Presutti, V.: Ontology design patterns. In: Staab, S., Studer, R. (eds.) Handbook on Ontologies. International Handbooks on Information Systems. Springer, Heidelberg (2009)
48. Office of Government Commerce. The Official Introduction to the ITIL Service Lifecycle, 2007th ed. The Stationery Office, London (2007)
49. Cannon, D., Wheeldon, D.: ITIL Service Operation, 2007th edn. The Stationary Office, London (2007)

Author Index

Printed in the United States
By Bookmasters